Industry Risk Communication Manual: Improving Dialogue with Communities

Billie Jo Hance
Research Associate
Environmental Communication
Research Program
New Jersey Agricultural
Experiment Station
Cook College, Rutgers University
New Brunswick, New Jersey

Caron Chess
Associate Director
Environmental Communication
Research Program
New Jersey Agricultural
Experiment Station
Cook College, Rutgers University
New Brunswick, New Jersey

Peter M. Sandman
Director
Environmental Communication
Research Program
New Jersey Agricultural
Experiment Station
Cook College, Rutgers University
New Brunswick, New Jersey

Lewis Publishers

Library of Congress Cataloging-in-Publication Data

Hance, Billie Jo.
   Industry risk communication manual: improving dialogue with communities
   /Billie Jo Hance, Caron Chess, Peter M. Sandman.
       p.   cm.
   ISBN 0-87371-274-9
   1. Risk communication. I. Chess, Caron. II. Sandman, Peter M. III. Title
T10.68.H36   1990
658.4′08—dc20   90-5974
                                                        CIP

This book represents information obtained from authentic and highly regarded sources. Reprinted material is quoted with permission, and sources are indicated. A wide variety of references are listed. Every reasonable effort has been made to give reliable data and information, but the author and the publisher cannot assume responsibility for the validity of all materials or for the consequences of their use.

Direct all inquiries to CRC Press, Inc., 2000 Corporate Blvd., N.W., Boca Raton, Florida, 33431.

© 1990 by Lewis Publishers
Second Printing, 1991

International Standard Book Number 0-87371-274-9

Library of Congress Card Number 90-5974
Printed in the United States

TABLE OF CONTENTS

ACKNOWLEDGMENTS

This manual would not have been possible without the assistance of many people who supported our efforts. We are particularly grateful to the Hazardous Substance Management Research Center at the New Jersey Institute of Technology, which saw the importance of producing an industry risk communication manual as quickly as possible and found a way to get funding to us in record time. A number of industry practitioners generously gave us substantial input concerning what the manual should cover and how it should do so: Fred Ellerbusch, Bristol-Myers Products; Kristin Elliott, Allied-Signal, Inc.; Keith Fulton, Exxon Chemical-Americas; Art Gebhardt, Manville Sales Corporation; John Slavick, Chemical Manufacturers Association; and Ben Woodhouse, Dow Chemical Company. We are also indebted to those people we interviewed who shared with us their insights and "war stories," used as examples throughout this volume. In addition, Ben Woodhouse pulled together a group of experienced risk communicators from Dow's Midland facilities for a spirited discussion that was extremely useful to the development of this manual. We also thank the Division of Science and Research of the New Jersey Department of Environmental Protection, which funded the development of *Improving Dialogue with Communities: A Risk Communication Manual for Government*, the basis for many of the concepts explored in this volume.

INTRODUCTION

In some respects, communication about environmental risks between industry and communities has made progress in recent years. Most recently, the Title III provisions of the Superfund Amendments and Reauthorization Act (SARA) have encouraged some companies not just to comply with the letter of the law, but to open up dialogue with the communities in which they reside. Although many companies and communities are still far apart in understanding, some are beginning to establish mutually respectful relationships and work toward common goals.

That's the good news. On the other hand, there are countless citizens across the country who do not have any idea what hazards exist in their communities and countless others who have little idea which hazards may be potentially serious. Industry communication efforts are by and large understaffed and underemphasized, and often communities have to fight for information that is metered out to them in tiny amounts — exactly what they ask for when they ask for it and no more. Many environmentalists view many industries as arrogant and paternalistic, and many industries feel many environmentalists are unreasonable.

Has there been any net progress? Most of the industry representatives and environmentalists who were interviewed for this manual think there has been and have provided examples of their own successes. In fact, interviews we conducted with government practitioners, environmentalists, and academics 3 years ago for a risk communication manual for government agencies centered around why risk communication is necessary.* This time, we found the "whys" had largely been estab-

---

* Hance, B. J., C. Chess, and P. M. Sandman, *Improving Dialogue with Communities: A Risk Communication Manual for Government* (Trenton, NJ: Department of Environmental Protection, January 1988).

lished, if not yet fully accepted. There are few who will now dispute the need to promote dialogue between a company and the community.

The questions this manual addresses, then, are "What kind of communication really makes a difference?" and "How do you do it?" Accordingly, this manual stresses the "hows" much more than the "whys": how to explain technical information in lay terms, how to be a better communicator, how to find ways to reach out to the community.

The suggestions in this manual are based on extensive interviewing, not empirical research. Because industry practitioners cannot wait until all the data are in, we are providing the best guidance we can. But more empirical research is desperately needed to "test" many of these suggestions, which should be considered "working hypotheses" rather than truths.

♦ WHERE THE GUIDELINES COME FROM

The guidelines, checklists, examples, and quotations that make up this manual come primarily from interviews with nearly 30 industry practitioners — and several environmentalists — across the country, as well as from our own experience training, advising, and consulting with many industries. In a few instances we have chosen not to reinvent the wheel, but rather to use or adapt material previously developed by us or someone else. (In these cases references are given so that the reader can get the material in its entirety if desired.)

Before the interviews were conducted, we consulted several industry leaders to (1) develop a better sense of what risk communication problems companies were finding the most difficult to deal with; (2) obtain names of industry practitioners who had a reputation for dealing with risk communication problems effectively and innovatively; and (3) solicit input about the most useful format for this manual. (As a result of this process, we decided not to cover such topics as crisis communication — which is amply covered by other literature — and communicating about specific situations — such as siting hazardous waste facilities or releasing Title III information.) We then conducted hour-long interviews with both community relations and technical people in several industries, as well as with several environmentalists. The goal of all interviews was to obtain risk communication approaches that were positive and creative. For the most part the communication methods presented here are currently being used by practitioners, but we also have included ideas that people had not yet implemented.

The inclusion in this manual of examples from a particular company's risk communication experiences does not imply our endorsement of that company's past, present, or future risk management practices — or even its other risk communication programs. Similarly, we are quite sure there are other companies we did not interview whose risk communication practices might appropriately be used as examples in this manual. While we tried to cast a wide net, we apologize to those whose programs we missed. Our goal was not to catalog all the effective risk communication programs, but rather to describe a limited number of useful examples. However, we suspect that there is room for improvement even among the best.

◆ HOW TO USE THE MANUAL

Each chapter in the manual covers a different aspect of risk communication. Chapter 1 covers the risk communication questions that many company managers are having particular trouble solving. This chapter lays the groundwork for the rest of the manual, and we suggest you read it before you read the rest.

Other chapters include the following:

- **Guidelines** — Each guideline is documented with rationale and examples.
- **Checklists** — Each guideline is followed by one or more pages of "how-to" checklists.
- **"Yes, buts..."** — The final section of each chapter responds to the most likely objections and concerns of readers.

We recognize that there is a tendency for busy people to skip directly to the checklist they need at the moment, so we have made each checklist as complete and self-explanatory as we can. However, we do not feel that effective risk communication can be achieved simply by checking items off a list, and we urge you to find some "non-crisis" time to read the entire manual.

◆ WHAT THIS MANUAL CAN AND CANNOT DO

This manual is for plant managers, public information specialists, technical staff, and other managers who need to deal with the public around environmental risk issues — or for those who guide people who do. We hope this manual will give you ready access to current thinking and practice on environmental risk communication. It can help you think through risk communication problems and plan risk communication strategies, and provide you with examples and background from other practitioners who have found themselves in similar situations.

Unfortunately, this book cannot provide you with a quick fix for your communication woes with community residents or environmentalists. It does not teach "environmental public relations," but rather how to plan and carry out two-way dialogue with those who are affected by your company's activities. No manual can change your company's policies and priorities regarding how you deal with the public. What the manual can do is provide you with a rationale for change and the encouragement of industry "success" stories.

◆ A FINAL WORD

We cannot stress enough the need to recognize that risk communication "successes" are incremental. Even if you conscientiously apply all the guidelines in this manual, it is still unlikely that you will be instantly accepted and loved by those who now reject and fight you. But you will see — as those quoted here have seen — real progress in your relationships with people outside your gates. And if you share those successes with others in your company, with management, and with colleagues in other companies, you may plant the ideas for future success stories.

LIST OF INTERVIEWEES

The following people are quoted and referred to throughout the manual.

June Andersen
Project Manager, Environmental Assessment
IBM Corporation
San Jose, California

Ken Brown
Director
New Jersey Environmental Federation
New Brunswick, New Jersey

Thomas A. Chizmadia
Director, Corporate Communications and Public Policy
CIBA-GEIGY Corporation
Ardsley, New York

Joan Ebzery
Director, Public Affairs
Clean Sites Inc.
Alexandria, Virginia

Thad Epps
Regional Director of Public Affairs
Union Carbide Corporation
South Charleston, West Virginia

Martin W. Ferris
Plant Manager
Air Products and Chemicals Inc.
Escambia, Florida

Scarlett Lee Foster
Environmental and Community Relations Manager
Monsanto Chemical Company
St. Louis, Missouri

Keith Fulton
Site Manager
Exxon Chemical-Americas
Baytown, Texas

Ray Kerby
Director, Environmental Programs
IBM Corporation
San Jose, California

Kelli Kukura
Senior Public Affairs Representative
Du Pont Corporation
Deepwater, New Jersey

H. F. (Bud) Lindner
Manager, Environmental, Health and Safety Operation
General Electric Silicones
Waterford, New York

Robert Manning
General Production Manager
Filtration and Minerals Division
Manville Sales Corporation
Lompoc, California

Donald McCambridge
Director, Human Resources Services
New Jersey Chamber of Commerce
Newark, New Jersey

Thomas McCollough
Refinery Manager
Sun Refining and Marketing Company
Tulsa, Oklahoma

Brent McGinnis
Director, Public Relations
Ashland Chemical
Columbus, Ohio

Brian McPeak
Manager, Public Affairs
Rohm and Haas Delaware Valley Inc.
Bristol, Pennsylvania

Fred Millar
Director, Toxic Chemicals Safety and Health Project
Friends of the Earth/Environmental Policy
    Institute/Oceanic Society
Washington, District of Columbia

Glenn S. Ruskin
Director, Government Affairs and Communications
CIBA-GEIGY Corporation
Toms River, New Jersey

Greg Schirm
Executive Director
Delaware Valley Toxics Coalition
Philadelphia, Pennsylvania

Roger Schrum
Manager, Media Relations
Ashland Oil Inc.
Ashland, Kentucky

Richard D. Stewart
Plant Manager
Du Pont Chambers Works
Deepwater, New Jersey

The following people from the Dow Chemical Company in Midland, Michigan, participated in a group interview.

Stanley L. S. Dombrowski
Manager, Environmental Affairs Group

Susan N. Dupree
Supervisor, Community Relations

Cindy C. Newman
Supervisor, Environmental Communications

Colin Park
Manager, Issues Management/Biostatistics

Douglas J. Robertson
Group Leader, Styron/Engineering Thermoplastics TS&D

Ben Woodhouse
Manager, Public Issues

1 MANAGERS' QUESTIONS ABOUT RISK COMMUNICATION

Although there is growing acceptance of risk communication principles, many companies still have reservations about applying the risk communication guidelines outlined in this manual. Specific reservations are addressed in the "Yes, but..." section that appears at the end of each chapter. This chapter addresses up front some of the broader questions practitioners have voiced about communicating with the public — good questions that deserve answers.

1. Why do people insist on zero risk?

Many people have lived with industries in their communities for years but haven't demanded zero risk. By and large, people insist on zero risk when they feel that they are being treated unfairly or that their concerns are being ignored — in short, when they are angry. If people suddenly seem to be demanding zero risk, companies have to understand the anger that motivates seemingly "unreasonable" demands and take responsibility for failing to reach out sooner. To avoid being confronted by community members who demand zero risk, it helps to involve them in decisions that affect them.

Another way of looking at the demand for zero risk, however, is to see it as a mandate for companies to work to reduce risks toward zero. Companies such as Monsanto and Du Pont have announced dramatic waste reduction policies that, while they do not promise zero risk, do give the concept some validity. Harold J. Corbett, senior vice president for environmental health and safety at Monsanto, describes a "typical chemical plant" 10 years from now: "It will have clean processes, with any necessary recycling...done at the point of production, so that no hazardous waste leaves the plant.... Process risk will be reduced to the point where serious accidents do not occur; accidental and routine emissions will be considered a relic of the stone age. The people living near the plant site will fully understand how the plant operates and what it manufactures."*

2. Why should I stir up trouble?

While Bhopal, the Exxon Valdez, and similar incidents have convinced many companies that they can no longer afford *not* to practice outreach to communities, other more fortunate companies may still be saying, in effect, "Why should I do outreach? People are basically happy with us." Or "They're not that interested. Besides, an accident like that can never happen here." But this attitude is a "time bomb," says Greg Schirm of the Delaware Valley Toxics Coalition, because there is no such thing as an "operation [with] no chance of any kind of accident." Only if your operation is perfect, says Schirm, might you want to "let sleeping dogs lie." The best time to build an honest relationship and get possible problems on the table is when the climate of opinion is positive — not when a crisis has made it hostile.

Being a "good corporate citizen" used to mean that a company provided jobs, volunteered time, and contributed to charities, says Thad Epps of Union Carbide. But that definition has changed: "Inside your gates...[is] not another world any more. What goes on in a chemical plant is of concern to a community.... You are perceived as a neighbor who could cause [the community] a lot of trouble.... You have to convince your neighbors that you are a responsible operator of your facility."

Fred Millar of the Friends of the Earth/Environmental Policy Institute/Oceanic Society says that even in communities where people seem apathetic, companies should still practice outreach. Says Millar, "It's just not honest to say that you have an apathetic community when you have in fact denied them and they don't even know that they have a right to get information [or] know that the information exists." In fact, says Millar, Title III is largely in response to communities' realization

* Corbett, H. J., "Chemicals and the Public," *Chem. Rep.*, May 30, 1988.

that they *do not* have the whole story. "Companies are willing to give us information about the probability of [an accident] happening, but not the *consequence*. They decide what risks are acceptable, but we can't because they haven't shown us the hazard information yet," says Millar.

3. Isn't it just outside activists who come in and stir up our communities?

Activists are often seen by companies as extremists who use communities for their own political purposes. And while many companies have a battery of specialists, both inside and outside the company, to advise them and represent their positions on various matters, they nonetheless find it difficult to accept communities' associations with groups that offer similar support.

In some cases the positions and tactics of an activist group may in fact be more extreme than those of local residents. But activists cannot create community concerns. They can only help communities to define and sharpen their concerns and give guidance on ways citizens can organize and be heard. Keith Fulton of Exxon Chemical, who has worked closely with environmentalists at the company's Baytown plant, says that although the environmentalists may not "represent the community as a whole, they do represent the concerns the average citizen, even a trusting one, may have in the back of his mind."

If you are concerned that an activist group is going to come into your community and "stir things up," you probably have not done as much outreach to the community — and to that group — as you should be doing. Scarlett Lee Foster of Monsanto says that "most of the time an outside group...may not have the community's best interests at heart. Then again, we may not have either, and there is a reason they can come in and get set up.... I think they get set up where people haven't established relationships."

Some companies who have dealt with national groups directly have learned to work for mutual respect and dialogue rather than agreement. Says Kelli Kukura of Du Pont of the company's meeting with Greenpeace, "...We really have the same goal in mind, it's just that ours is a day-to-day, down-in-the-trenches goal for source reduction and waste minimization. They admitted theirs is a pie-in-the-sky viewpoint and they're holding it up for industry to try to meet."

4.   No matter what we do, it's never enough. Should we bother?

Companies sometimes feel that environmentalists — and many communities — are too demanding, that regardless of the extent of companies' efforts, environmentalists always want more. It is true that outreach often leads to more community demands as communities learn what makes sense to demand. Companies whose goal is simply to appease communities or to win them over rather than to listen and respond to community concerns will find that what they do *will* never be enough.

Environmentalists say that companies often give mixed messages or promise far more than they are in fact willing to deliver. For example, says Greg Schirm of the Delaware Valley Toxics Coalition, a Philadelphia hospital contacted local groups to put together a community relations committee to help the hospital develop plans for a proposed incinerator on the site. While the committee helped to define oversight rights, access to information, and what would constitute violations, the hospital brought in consulting engineers "to explain why there would be no risk...[and] that people wouldn't have much to worry about." Says Schirm, "On the one hand [the hospital] admitted there was some kind of risk in these operations...but when it came down to potential problems, they didn't really want to talk about anything other than how nothing is going to happen.... I think that makes people uneasy.... If they're trying to work with the community and be open and straightforward about these things, then they should be clear about what happens in the event of a problem."

Ken Brown of the New Jersey Environmental Federation feels that companies often assume that their scientists know "the truth" and that if the community will simply listen to the company's experts, the community will understand and agree. The problem with this, says Brown, is that it doesn't "give real legitimacy to community concerns, and [the company] is not willing to give up any power...."

Ben Woodhouse of Dow Chemical says that his company's outreach to activists was largely ineffective until Dow had something to offer them. Says Woodhouse, "We started out in a formalized fashion. We drew up a list of 25 groups that didn't like us the most.... The problem was that we were going in there trying to get to know each other and didn't have anything specific to put on the table. So we backed off from that somewhat formalized and cumbersome outreach program to being very targeted. We start out with our issue. We make a list of the critics, and decide which ones we want to go for on that issue."

5. How can I do outreach with limited resources?

While it is true that effective communication is unlikely to happen if resources are not devoted to it, those who have successful outreach programs say that even more important is the *commitment* to communicate. One significant difference between those who have outreach programs and those who do not is how they prioritize communication tasks among their other functions (see page 129). For most companies who are doing community outreach, the key to finding the time to do it is to see it not as an "add-on," but rather as part of their operational tasks. Kelli Kukura of Du Pont says that for the company's plant managers, communication is "considered part of the job" and written into job descriptions. Other companies report management directives that stress the importance of outreach to the community. According to Keith Fulton of Exxon, "We take on added workload continuously in companies...it's the nature of change. But when it's internally driven we don't say 'I can't do it. I don't have the resources.' We just adjust."

According to Don McCambridge of the New Jersey Chamber of Commerce, some very small companies may feel that doing outreach to the public is simply beyond their means. "They are not necessarily ignoring...but just rearranging priorities. There's so much to do.... They can't delegate responsibility because they're not making enough profit to hire the staff," says McCambridge. But Greg Schirm feels that "if a company is handling dangerous chemicals and it can't estimate its releases and figure out what the actual effects in the community might be...they're too small to be in the business."

6. How can we overcome the alarm created by the media?

Unfortunately, the media don't have to dig too deep or embellish the facts to find frightening environmental accidents and issues to report on. While the media do sensationalize, they do so much less when companies practice proactive communication than when they stonewall — and it is not sensationalizing to cover a local controversy or accident. Companies are far better off being responsive to the media, recognizing deadlines and other constraints, and establishing relationships with reporters than complaining about unfair media coverage (see Chapter 5).

Like environmentalists, the media may amplify community concerns, but they do not create them. "We have no right to ask for favorable news coverage," says Glen Ruskin of CIBA-GEIGY. "The thing we have to strive for is balance and that the stories are fair."

7. Can risk communication help us to avoid conflict with communities?

It depends. If you think risk communication is a way to get communities to see things the way you do, then it probably won't help much. The conflicts that will arise when people feel you are trying to paper over their concerns with public relations techniques may be worse than those with which you are currently dealing.

Avoiding conflict, in fact, may not even be the point. Companies that have made progress in their communication with communities do not claim that it is always smooth sailing. But, according to Keith Fulton of Exxon, even though open communication is often frustrating, "the positive side of this is if we can show that we can just give some power, share with citizens, find out what their concerns are, and hopefully work together, that this will be an advantage." Kelli Kukura of Du Pont says of working with a local citizens group, "We understand their position and they understand ours, but we just can't come to a middle ground. But at least we're sharing information. We've gotten that far."

8. Why do we have to work so much harder than government or environmentalists to get people to trust us?

Industry often feels that it alone has its feet to the fire, that everything it does is scrutinized by the public and the media. It is true that people demand higher standards from those who have power over them; companies have people's lives in their hands and therefore must prove themselves to be *at least* as responsible as the environmentalists who represent people's concerns. While this may seem unfair, it is also justified by history (industry has in the past misled the public), by position (industry has a reason to mislead the public), and by power (industry has the power to do damage).

Ray Kerby of IBM Corporation says the government's lowered credibility puts more pressure on industry. "Mechanisms [for dialogue] are set up in government that aren't used because government sometimes has no credibility." However, it does industry no good to undercut the institutions that people must trust. Kerby feels that "there has to be some even-handed processes and some expectations we can all depend on both for public participation and industry performance."

9. Why don't people understand and accept the technical evidence?

Industry representatives often feel that if people could just understand the technical information, then they would agree with industry and accept the risks industry wants them to accept. But this ignores other nontechnical components of risk — for example, fairness, control, etc. (see Chapter 2). These aspects of risk are a legitimate part of the issue which companies often neglect just as they claim the public neglects the technical side.

Even if you focus only on technical issues, the evidence of low risk often isn't as strong as companies make it out to be and there is a fair amount of disagreement even among technical people. Companies rarely invest enough time and other resources generating the data (much less supplying and explaining the data so people can understand). People are being asked to trust that industry knows everything it needs to know and is telling them everything *they* need to know. Not trusting that is not the same thing as rejecting the evidence.

Still, it is often true that people *do* reject the evidence — especially when they feel ignored or mistreated. The answer isn't to talk louder or even more clearly. It is to *listen* better.

10. Why should we go beyond complying with the regulations?

The best reason to go beyond compliance is because doing more benefits everyone. Compliance with the laws is the minimum that companies have to do — communities have every right to want more. As Ken Brown of the New Jersey Environmental Federation says, "In many instances of source and waste reduction, companies have gone over and above the law." In order to achieve meaningful reductions, companies have to take the lead in forcing technology issues. Going beyond the minimum also helps a company get "ahead of the curve" of coming environmental regulations. And if your company is doing more than simply complying with the law, taking the next step to talk to people about your efforts makes good sense.

11. Won't communicating openly about what goes on at our plant make us more vulnerable legally?

In March of 1988, the Chemical Manufacturers Association's General Counsel issued an advisory that questioned companies' "traditional legal response" of providing "the minimum amount of data." In light of new legislation requiring disclosure of data "that might be misunderstood and misinterpreted," says the report, "it may be wise to set the record straight *before* confusion and concern overwhelm the situation. Providing additional information about the chemicals in question and the safety practices employed by a facility...may persuade skeptical members of the public...that they need not be concerned." The report admits that this type of disclosure may be risky, "but it may be a risk worth taking," because the public's opposition may create far more problems for a company than most lawsuits.*

Thad Epps admits that the objectives of the company's legal and public relations departments are often "at opposite ends.... The lawyer's concern is that if you admit to something, that puts you at a tremendous disadvantage." However, says Epps, at Union Carbide they try to "understand each other's needs, working together and touching base with each other."

12. How can I deal with the community when politicians are constantly changing the agendas and going on crusades?

There is no question that environmental issues are — and should be — fodder for candidates' election campaigns. However, politicians are not the only ones with agendas, and pointing the finger at someone for pushing an issue simply to get elected may not only be unfair, it also

---

* Hogan and Hartson, "General Counsel's Advisory Supplement: Managing the Release of Environmental Information" (Chemical Manufacturers Association, March 1988).

can backfire. Attacking the credibility of anyone with whom you disagree can put you in a lose-lose situation. It makes more sense to do your homework, look at the statements being made by a candidate, clarify your own position on the issue, and try to create a dialogue with that person or his or her representatives.

2 RISK COMMUNICATION FUNDAMENTALS

♦ **RECOGNIZE THAT EFFECTIVE RISK COMMUNICATION DEPENDS LARGELY ON EFFECTIVE RISK MANAGEMENT**

The current emphasis on risk communication has led some to believe that there is a new way to talk about risk that might make people accept risks they now find unacceptable. However, simply talking openly and honestly about risks, although important progress for companies to make, is not enough — companies have to *do* something about them. Those industry practitioners who have had successful interactions with communities have succeeded because of their willingness to back up words with actions. As Colin Park, manager of Issues Management/ Biostatistics for Dow Chemical, points out, "They don't want to hear what the numbers are or the comparison.... They want to hear what you're doing to get rid of it."

Scarlett Lee Foster of the Monsanto Chemical Company feels that Monsanto's 1988 commitment to reduce air emissions by 90% world- wide by 1992 helped her to communicate the company's Title III air data, by showing that the company was taking seriously public con- cerns about chemical emissions. This goes a lot further, says Foster, than coming up with good communication strategies. "Risk communi- cation is only as good as your performance. Where I have been success- ful, Monsanto has been doing the right thing." Foster adds that risk communication does not provide a way to paper over ineffective risk management. "I can't make a bad situation good with just communica- tion.... What really works is to say 'I recognize I've got a risk. This is what I'm going to do to fix it.'"

Although companies may not be ready to make a global commitment to waste reduction as Monsanto has done, many companies are realizing that they must respond to the public's environmental concerns. H.F. (Bud) Lindner of General Electric Silicones says that when he could not answer a television reporter's question about how GE would improve its performance after a spill into a river just upstream of a local waterworks, he realized that the company needed to install a monitoring and containment treatment system, that there was no way to communicate away the need to do something about the problems at the plant.

⮑ Improvements Communities Often Request

1. Reductions in current emissions over past performance.
2. Commitment to reduce emissions even further in the future.
3. Installation of control equipment.
4. Additional training of staff in safety and emergency response.
5. Assistance in improving community emergency response by training police and fire departments, donating needed equipment, etc.
6. Improvements to decrease the likelihood of an accident occurring.
7. Control measures to decrease the magnitude of an accident if one were to occur.
8. Thorough emergency response plans.
9. Air and water monitoring.
10. Processes that minimize waste at the source.
11. Reduction in the amount of hazardous raw materials stored on site.
12. Mitigation and compensation for damage already done, and policies to redress harm from future damage.
13. Existing information on health risks.
14. New environmental and health studies for on- and off-site risk.
15. Access to data so that the community can verify the improvements.
16. Inspection of the facility, accompanied by the group's technical advisor, to verify the company's improvements.

♦ PAY ATTENTION TO OUTRAGE FACTORS

Industry representatives are often confused and frustrated by public reactions to risk. They find themselves in battles with people about risks that may cause fewer than one in a million increased cancer deaths. Yet these same people may smoke or drive without seat belts. When confronted with the apparent contradiction, people become even angrier.

In fact, there is growing evidence to suggest that the risks that frighten people may not be the same ones that kill them. While companies may focus on hazard evaluations, risk assessments, and monitoring, communities can be equally concerned about many factors besides the data.

Part of the problem is definition. To scientific experts and industry, risk means expected annual mortality. But to the public (and even the experts when they go home at night), risk means more than that. Let's redefine terms. Call the death rate (what the experts mean by risk) "hazard." Call all the other factors, collectively, "outrage." Risk, then, is the sum of hazard and outrage. The public pays too little attention to hazard; the experts (and too often industry) pay too little attention to outrage. While industry certainly should work to make the hazard data available and easy to understand, companies need to work equally hard at dealing with the other dimensions of people's concerns. Ignoring these other dimensions — or labeling them irrational — is virtually guaranteed to raise the level of hostility.

Companies can get into battles if they stubbornly assert their own "rightness," rather than dealing with community outrage. Scarlett Lee Foster of Monsanto tells how the company's handling of some citizens' complaints about odor was based on management's desire not to give in when they felt they were right—rather than genuinely trying to deal with the concerns of community members. When two women repeatedly complained of odors associated with a Monsanto process, the company got into "an adversarial relationship with them.... It was a battle: Whose nose is better?" While the company focused on the risk, the community was as concerned about an issue other than the risk — odors. Rather than do air monitoring, as the citizens requested, Monsanto basically asked the citizens to believe the company that there was nothing wrong. Foster says she now realizes that "we should have shown them the unit. We should have explained better...why there is blue smoke...why we didn't think there was an odor associated with it." Monsanto might also have been better off doing the air monitoring the community requested, even if management felt there was no problem.

Why were citizens so upset when the company saw so little to be concerned about? Not only were the citizens offended by the odors (whether they were risky or not), but whatever risk did exist (no matter how small) felt even riskier because it was imposed by the company and citizens felt very little control.

✔ Components of Community Outrage

When experts examine a risk, they focus almost exclusively on mortality and morbidity — how many people are likely to die or get sick as a result. But citizens define risk much more broadly, considering a wide range of "outrage factors" in determining how risky they consider the situation at hand.* As you examine the factors below, remember that they are *not* "misperceptions" of the technical data; rather, they are aspects of the risk quite apart from the data that most people consider important and relevant. It is therefore a mistake to dismiss these factors as "just outrage." Working to keep outrage low is as much a part of risk management as working to prevent mortality and morbidity.

1. **Voluntary vs involuntary** — People feel much less at risk when the choice is theirs. Consider the difference between getting pushed down a mountain on slippery sticks and deciding to go skiing. Find ways to share decisions with concerned residents.

2. **Natural vs industrial** — A natural risk like a flood is midway between a voluntary and a coerced risk; we are all more forgiving of "God's coercion" than of corporate coercion. You can't make industrial risks into natural ones, but you can avoid comparing your effluent to peanuts.

3. **Fair vs unfair** — Even if a situation entails more benefits than risks, the people who bear most of the risks often reap little of the benefits. This unfairness naturally provokes outrage. Reduce risks where you can — then compensate where you cannot reduce. To find out what sort of compensation is most appropriate, ask the community.

4. **Familiar vs exotic** — Familiar risks and familiar surroundings diminish outrage. That's why homeowners don't much fear radon and plant workers sometimes don't pay enough attention to safety rules. Facility tours, mall displays, school programs,

---

* The definition of these variables is the result of years of empirical research by social scientists, particularly the groundbreaking work of Baruch Fischhoff, Paul Slovic, and Sarah Lichtenstein.

and the like can increase familiarity and reduce outrage — especially if they don't evade the tough issues.

5. **Not memorable vs memorable** — Whether through personal experience or media experience, memorable incidents and images of risk exacerbate outrage. When such incidents and images are present in people's minds, ignoring them just makes the problem worse. Discuss them *before* you are accused of them.

6. **Not dread vs dread** — Some diseases are more dread than others (cancer versus asthma, for example); so are some exposure paths (water versus air) and some hazard categories (waste versus product). Since you can't reduce dread, your best bet is to acknowledge it.

7. **"Knowable" vs "unknowable"** — Do the experts agree? Do they seem to understand the hazard? How big are the error bars in the risk estimate? Is the hazard visible or otherwise detectable? Often you can increase knowability by giving citizens access to the data — put a satellite thermometer on top of an incinerator, for example, so citizens know that it is burning hot enough.

8. **Morally irrelevant vs morally relevant** — To many, pollution isn't just harmful; it's wrong. Its moral relevance makes the language of cost-risk and benefit-risk tradeoffs seem callous — like a police chief discussing how many child molestations are "acceptable." Like the chief, you don't have to get to zero, but you must take zero seriously as a goal.

Three other outrage factors are so important that we deal with them specifically in other guidelines in the manual.

9. **Controlled by the individual vs controlled by the system** — Most people feel safer driving than riding shotgun. Being at the mercy of someone else provokes the most outrage when "someone else" is a faceless corporation. Techniques to share control range from informal surveys that ask people their concerns to formal advisory committees (see pages 27 to 34).

10. **Trustworthy vs untrustworthy** — Polluting industries are widely distrusted, and people make the conservative assump-

tion that an untrustworthy company might well be dangerous as well. Over the long haul, you need to build more trust (see pages 37 to 42). Meanwhile, you need to demand less trust by making your actions public, collaborative, and accountable.

11. **Open process vs closed process** — What sort of relationship has your company built with the community? Do you admit past errors or deny or ignore them, release information promptly or hide it (see pages 56 to 59)? How do you deal with people's concerns — courteously and responsively, or with arrogance, defensiveness, or techno-babble (see pages 49 to 55 and 73 to 77)?

◆ INVOLVE PEOPLE IN ONGOING RISK DECISIONS

Many companies are beginning to come to grips with what it means to involve citizens in environmental decisions that affect them. While some have invited citizens into their plants for discussions on plant improvements, others are more cautious and even resistant to acknowledging the value of citizen input or oversight. One reason for this resistance is fear of having citizens who don't understand the "whole picture" trying to push companies for unrealistic solutions to problems. Another fear is loss of control. Ben Woodhouse of Dow Chemical says, "It does tie your gut up in some knots when you think about that kind of thing.... We're evolving into it.... We are at the point where we recognize that they [communities] need to participate in our thought processes, the planned development of what we need to do to make our operation better...."

While many companies are still fearful about the notion of sharing some measure of decision-making with local citizens, environmentalists see citizen oversight as benefiting both the community and the company. According to Fred Millar of the Friends of the Earth/Environmental Policy Institute/Oceanic Society, if the community is involved in oversight activities, it is able to "trust, but verify" the company's progress toward its stated environmental goals. Greg Schirm of the Delaware Valley Toxics Coalition says that the benefits to the company may be greater than companies expect: "The more people know about an operation, the more they make suggestions...and to the extent that management agrees with some of those things and makes some effort to improve the situation, then I think it's in the company's interest.... Not to do that means there's always a potential...for an outraged community saying 'Get this plant out of our community.'" Schirm goes on to say that there is also the danger that citizens can go to the government with their complaints and try to "force the whole situation into court," resulting in long, expensive legal battles. Even more to the point, these types of battles perpetuate strained relations between a company and the community in which it resides.

One way companies can involve community groups is through plant inspections. Ken Brown of the New Jersey Environmental Federation says that the Dynasil Corporation in New Jersey agreed to have a community group (Coalition Against Toxics) tour the plant with its technical advisor. As a result of that inspection, says Brown, the group presented to the company a report containing several recommendations that the company subsequently implemented. Brown characterizes the meetings between the group and Dynasil as a "very cooperative, open exchange of information."

Martin Saepoff, president of the Dynasil Corporation, acknowledges that companies are fearful of what they feel are unreasonable demands from communities. He says the key to the success of this interaction was sincerity on both sides: "I could see where a lot of management in industry could be concerned about nosy neighbors coming in and trying to stir up trouble.... When I became assured that these were sincere people who were simply concerned about some very straightforward matters regarding their environment, I wanted them to realize that we were sincere and not trying to get away with anything...."*

Citizen demands will not always look as reasonable as in the previous example. However, even when people are demonstrably angry and request changes that companies find difficult to make, that does not mean that their concerns are insincere or their requests invalid. Keith Fulton of Exxon Chemical has been working with a community group and expresses frustration at some of its requests as well as the time and effort that has gone into working with the group. "It has been very time-consuming and frustrating," says Fulton, "and essentially we've been working with six to twelve citizens in a community of 60,000 people." Fulton cites hours of weekend meetings to go over information that he feels is not relevant to most citizens' concerns. However, Fulton feels it is important to work with *all* the people who voice concerns — both the "reasonable" ones and those who are "causing a big stink and yelling and screaming."

The day-to-day operations of a company rarely involve much interaction with the public. But increasing the level of public participation makes sense for at least four reasons: to avoid (or reduce) controversy; to respond to citizen requests for participation; to have a chance to

* From Chess, C., S. K. Long, and P. M. Sandman, "Making Technical Assistance Grants Work," (New Brunswick, NJ: Environmental Communication Research Program, April 1990).

respond to citizen concerns and misimpressions; and to get the benefit of citizen suggestions.

Citizen participation is not an on-or-off phenomenon. Rather, there are many "steps," as the "Ladder of Citizen Participation" in Figure 1 illustrates. Note that the "Inform" and "Consult Pro Forma" steps may feel like community input to the company, but they may not feel that way to the community. Some interactions should reach at least the "Consult Meaningfully" level. Above all, you should know what level of interaction you want and make that clear to the public. In particular, avoid promising one level and delivering a lower one.

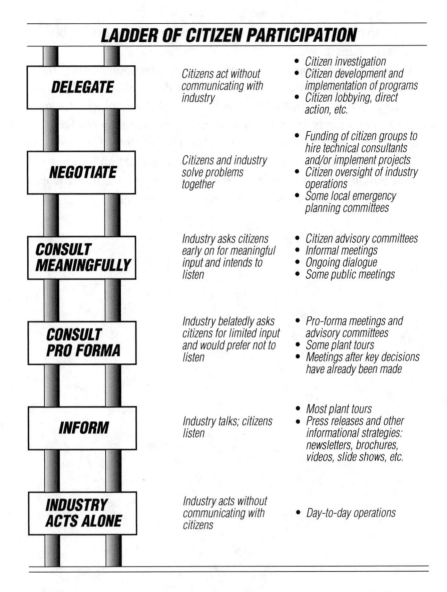

LADDER OF CITIZEN PARTICIPATION

DELEGATE

Citizens act without communicating with industry

- Citizen investigation
- Citizen development and implementation of programs
- Citizen lobbying, direct action, etc.

NEGOTIATE

Citizens and industry solve problems together

- Funding of citizen groups to hire technical consultants and/or implement projects
- Citizen oversight of industry operations
- Some local emergency planning committees

CONSULT MEANINGFULLY

Industry asks citizens early on for meaningful input and intends to listen

- Citizen advisory committees
- Informal meetings
- Ongoing dialogue
- Some public meetings

CONSULT PRO FORMA

Industry belatedly asks citizens for limited input and would prefer not to listen

- Pro-forma meetings and advisory committees
- Some plant tours
- Meetings after key decisions have already been made

INFORM

Industry talks; citizens listen

- Most plant tours
- Press releases and other informational strategies: newsletters, brochures, videos, slide shows, etc.

INDUSTRY ACTS ALONE

Industry acts without communicating with citizens

- Day-to-day operations

FIGURE 1. Ladder of citizen participation. Adapted from Hance, B. J., C. Chess, and P. M. Sandman, *Improving Dialogue with Communities: A Risk Communication Manual for Government* (Trenton, NJ: Department of Environmental Protection, January 1988). The ladder of citizen participation is derived, in part, from Arnstein, S., "A Ladder of Citizen Participation," in *The Politics of Technology* (1977), 240.

↙ Citizen Advisory Committees

Citizen advisory committees are not a new idea. Many companies have formed one at some point. However, too often the committee formed around a specific issue incorporates corporate allies or particular interests and excludes others. The committee can thus wind up being more of a buffer between the company and the community than an amplifier for community concerns.

The Rohm and Haas facility in Bristol, Pennsylvania has formed an advisory committee that, while not solving all the problems inherent in advisory committees, is moving toward becoming an integral part of the company's relationship with the surrounding community. Some of the ways in which this committee distinguishes itself from other such groups are the following:

- Extensive networking was used to find interested parties when the committee was formed and continues on an ongoing basis to ensure that the group is representative of the surrounding community.
- The committee has three primary "ground rules": (1) members are not bound by secrecy, but are free to tell those outside the committee anything they hear inside; (2) there are no boundaries to the issues the committee will explore, including corporate responsibility, future projections, emergency planning, etc.; and (3) there are no votes, that is, the company does not ask the committee to be held accountable for company actions.
- The company, acknowledging that such a group has internal issues that are distinct from company issues, has paid for an organizational consultant to help the committee with its own internal organization.
- At the company's expense, the committee places advertisements in local newspapers that discuss a range of issues affecting the community. The advertisements invite community response directly to committee members or to a post office box available only to committee members.
- Both company and community representatives work their schedules around the meetings. To promote continuity, there are no designated alternates.

Brian McPeak, manager of public affairs at Rohm and Haas, has been working with the group since it was formed in 1986. He says one of the most significant contributions of the committee is the way it has changed how the company makes decisions: "We have become more sensitive and aware as a company to the impact of the operation and to community concerns. Internally, it forces us to better evaluate our options [because] we have to lay out for them our thought process. When you're committed to that, you're committed to sharing with them the options you have. It then makes us as an organization more carefully evaluate all the options."

✔ Suggestions for Working with Activists*

Activist organizations deserve special attention in the risk communication area. Every strategic public affairs or community outreach plan should address the issues involved in communicating with activist audiences. The following questions may help you frame an approach to the issues:

1. What activist groups exist in the local area? In the state?
2. What local groups are linked to national organizations (e.g., Natural Resources Defense Council, Environmental Defense Fund, Greenpeace, Sierra Club, Conservation Foundation, etc.)?
3. Will our facility make an effort to reach out to recognized local activist groups, or will we wait until we are publicly targeted on a special issue?
4. Has the facility had previous encounters with activist groups? What was the outcome? How would we approach it if we had a second chance?
5. Are employees involved with these organizations? Neighbors?
6. Does your company or your plant contribute to any activist group or support university activities linked to public interest groups?
7. Are there community activities or projects the plant could sponsor jointly with local activist groups to help build communication channels and trust?
8. Does your facility have a response mechanism (security, media relations, etc.) in place that can be activated when you are targeted by an activist group?
9. Do you have "third party support" for your facility and its health, safety, and environmental protection record?
10. Are there benefits to be gained from management joining local activist groups?

---

* Chemical Manufacturers Association, *Communicating Risk: The CMA Workshop* (Washington, DC, 1989). This document was prepared for CMA by Rowan and Blewitt, Inc., and the Center for Risk Communication at Columbia University.

Yes, but...

There is no reason for the distrust people feel toward us and no reason we should have to deal with unfounded fears.

The enormous distrust can be very painful to those who feel they have worked hard for years to run environmentally sound plants. The public tends to remember failures over successes, slights over kindnesses.

On the other hand, industry's record in the United States is not without blemishes. Environmental groups can point to a variety of abuses, including industry secrecy. Superfund sites are not a figment of the country's collective imagination. Communities will have difficulty at times sorting out your track record from the headlines they have read for years about misdeeds across the country.

Instead of resenting the distrust, industries may need to work twice as hard at acting trustworthy and accept that trust will take time to build. In the meantime, leaning heavily on a trust that isn't there — demanding to be trusted — is sure to backfire.

We just want to educate people so they understand what we are doing. If they understood the data as we do, they would accept the outcome.

These statements may seem noncontroversial, but to those outside of industry they smack of arrogance and "industry propaganda." Assuming that those you are talking to are misguided and that you hold a monopoly on truth rarely wins respect, let alone converts.

Although it is true that you may have more data, the public has interests and concerns that go beyond the data. Dismissing those concerns — even if they seem misplaced — will more often than not lead people to dismiss your data.

Our job is running safe plants, not dealing with the public's emotions.

Making plants safe and environmentally responsible is certainly a primary goal for those inside and outside the plant. However, management options are rarely perfect. In fact, concerns raised by those outside the plant can lead to better technical solutions. Unfortunately, there have been many times when companies dismissed public concerns that then became real, or ignored public suggestions that could have helped.

Perhaps just as important, issues such as fairness, trust, and control are issues that do matter — and should matter. Outrage is not just a distraction from the "important" issues, but is important in its own right.

Finally, ignoring issues about which people care deeply (because your data suggest they shouldn't care deeply) is virtually guaranteed to increase rather than decrease their concerns while making you seem uncaring.

3 THE PROCESS OF RISK COMMUNICATION

♦ **ACTIVELY SEEK TO IMPROVE YOUR TRUST AND CREDIBILITY BY PAYING ATTENTION TO PROCESS**

If you want credibility you have to "be credible."* Companies bemoan their loss of trust and credibility with the public, but fail to recognize how their day-to-day actions affect their trust and credibility. "Being credible," most experts and practitioners agree, involves more than being scientifically correct, or being a "good corporate citizen" in the traditional sense of civic responsibility (see page 10). As important — often more important — is having a credible process, which includes not just being responsible, but also being *responsive* to the community. While all the guidelines in this manual are in some way related to achieving more trust and credibility for you and your company, the key is simply to be aware of how what you do — your process — affects the way you are seen.

Citizens often complain that both government and industry don't talk to them and keep them apprised of developments. Joan Ebzery of Clean Sites, Inc. says that the company's recognition of the importance of process when it was involved in cleaning up PCBs in Holden, Missouri, prompted company officials to have very close contact with the community: "Every time we had to leave the site for a while we tried to let people know.... We put six or seven fact sheets out...had two

---

* James Callaghan in Hance, B. J., C. Chess, and P. M. Sandman, *Improving Dialogue with Communities: A Risk Communication Manual for Government* (Trenton, NJ: Department of Environmental Protection, January 1988).

or three town meetings.... We constantly felt the need to explain to the community what was going on."

Companies often ask communities to trust them, rather than building a trustworthy process. Keith Fulton of Exxon Chemical tells two stories that illustrate how a company can build trust instead of demanding trust. When a citizen at a meeting complained that there was black material on his roof that he suspected came from Exxon Chemical, Fulton offered to come out and take a sample of the material. On another occasion, when a woman complained that the Baytown facility was killing town trees, Fulton hired a tree expert who found the trees were suffering from drought. When the woman then questioned the company-hired tree expert's opinion, Exxon paid for the woman to hire her own expert, who said the same thing. While the woman, says Fulton, is "no friend of ours, at least she's convinced that we aren't killing the trees...."

It is important not to underestimate how being responsive in small, day-to-day ways can help a company's credibility. For example, Air Products and Chemicals in Pensacola, Florida, has a 24-hour hotline for complaints, and calls are dealt with immediately regardless of the time of day. Plant manager Martin Ferris cites an instance in which a person called about a black cloud. The responder went out and climbed a stack to see what it was and got back to her immediately about the source of the cloud (which was from agricultural burning, not from the plant).

✔ What to Do When Trust is Low

Several of the companies interviewed for this manual have had to deal with situations of extremely low trust, usually in response to a spill, leak, or accident (for example, Union Carbide after Bhopal, IBM after groundwater pollution in California) or after proof of health impacts from a company's products (the Manville Sales Corporation and asbestos). Here are some suggestions for dealing with such situations from those who have been there.

1. **Don't try to ignore or cover up the mistake. Acknowledge it** — It won't go away any time soon and will probably always be in the collective public memory. If you deal with it up front, you will avoid having to respond defensively about it.

2. **Don't get angry or defensive when "past sins" come up long after the particular events** — Simply recognize that they will come up and concentrate on improving your record. Better yet, bring them up before someone else raises them.

3. **Say what you are doing to deal with the mistake and prevent it from occurring again** — For example, Union Carbide's plant in Institute, West Virginia, similar to the unit in Bhopal, was shut down for 4 months after the Bhopal leak while the company made safety modifications.

4. **Don't try to weasel out by saying it was a different plant, different management, different regulatory climate, etc.** — First take responsibility and acknowledge your accountability. Then talk about what you have done since then to make such an accident less likely.

5. **Show that you have learned from your mistake** — For example, Robert Manning of the Manville Sales Corporation says the company's problems with asbestos have forced it to pay more attention to environmental and occupational health issues than most industries. While this cannot erase the company's past behavior, it does help illustrate its commitment to different behavior in the future.

6. **In some situations in which trust is extremely low, it may help to begin trying to communicate with those community organizations that are willing to talk with you — but not as**

a way to avoid the "tough groups". It's important to deal directly with people who have concerns about your facility — For example, IBM in the Santa Clara Valley, California, continues to be at serious odds with a key representative of a local environmental group over groundwater problems there, but is talking with local government, industry, regulatory agencies, and other segments of the public, such as the League of Women Voters.

7. **Expect to try twice as hard** — The burden is on you.

⤸ Ten Ways to Lose Trust and Credibility*

Take a good look at most risk communication "horror stories" and you'll probably find a major breakdown in trust between the company and the community. The next time someone comes to you with a sob story about communicating with the public, you might want to hand him or her this tongue-in-cheek list; or better yet, hand it out before the damage is done.

1. **Don't involve people in decisions that directly affect their lives** — Then act defensive when your policies are challenged.
2. **Hold onto information until people are screaming for it** — While they are waiting, don't tell them when they will get it. Just say, "These things take time," or "It's going through quality assurance."
3. **Ignore people's feelings** — Better yet, say their feelings are irrelevant and irrational. It helps to add that you can't understand why they are overreacting to such a small risk.
4. **Don't follow up** — Put returning phone calls from citizens at the bottom of your "to do" list. Delay sending out the information you promised people at the public meeting.
5. **If you make a mistake, deny it** — Never admit you were wrong.
6. **If you don't know the answers, fake it** — Never say "I don't know."
7. **Don't speak plain English** — When explaining technical information, use professional jargon. Or simplify so completely that you leave out important information. Better yet, throw up your hands and say, "You people could not possibly understand this stuff."
8. **Present yourself as a slick corporate apologist** — Wear a three-piece suit to a town meeting at the local grange, and sit up on stage with seven of your colleagues who are dressed similarly.

---

* Adapted from Chess, C., B. J. Hance, and P. M. Sandman, *Improving Dialogue with Communities: A Short Guide for Government Risk Communication* (Trenton, NJ: Department of Environmental Protection, January 1988).

9. **Delay talking to other companies and government agencies involved** — Or other people involved within your own company — so the message the public gets can be as confusing as possible.

10. **If one of your scientists has trouble relating to people, hates to do it, and has begged not to, send him or her out anyway** — It's good experience.

◆ IDENTIFY AUDIENCES AND ADDRESS ALL THE KEY ONES

"The public" is not one homogeneous unit, but rather made up of many smaller "publics." Most companies are aware of their more obvious publics — employees, residents whose homes are directly adjacent to the facility, local officials, and the media. At the same time, they often err either by failing to recognize more obscure audiences (for example, people in outlying communities, or those who have not previously been concerned) or by avoiding other audiences (for example, environmentalists). Instead, companies should make a point of identifying the publics they have forgotten and attempting to contact the publics they least want to talk to. Many seemingly smooth communication efforts have been suddenly derailed by the emergence of a public that was not previously recognized or was purposely excluded.

When Ashland Chemical wanted to site a new distribution facility in Greensboro, North Carolina, the company "did not go broadly to the community," but had meetings with the county elected officials, according to public relations director Brent McGinnis. These meetings went well, and the company was "continually encouraged." However, midway through the process, says McGinnis, citizens of the neighboring community of Highpoint learned of the proposed facility and became "quite irate" because the site was within the Highpoint watershed and they hadn't been consulted. These citizens mobilized into opposition to the facility, and, although Ashland's subsequent public information campaign has had some positive results locally, there is still opposition. Says McGinnis, "Where we came up short is in simply not having identified everybody who might have an interest." McGinnis adds that the idea of going out to a community with which the company has no "track record" is frightening to management, which feels that it might be shot down as people react simply "to chemicals rather than to Ashland." However, as this example illustrates, it is far riskier to leave a group out of your communication efforts.

Identifying audiences is not a one-shot effort, but must be done continually, with regular updating and efforts to seek new ones. Rohm and Haas went to great lengths to identify all interested parties for its advisory board (also discussed on page 31), but failed to recognize that a previously uninvolved neighborhood's population had been slowly evolving to a younger, more concerned group of residents — a "new base in the community" that was not represented on the board. When the company's proposed incinerator "pushed them to the brink," Rohm and Haas was forced to recognize it "didn't understand the dynamics of the community." Says Brian McPeak, "We have to make sure we are bringing in the changes that take place in the community."

⟋ Identifying Audiences*

Identifying audiences is no more than thinking through very specifically who might want to be talking to you. Although you may feel like avoiding groups or individuals who may be difficult to deal with, these are often the most likely to raise objections if they are not consulted early. In fact, if you would prefer not to hold a dialogue with a group because it is hostile or otherwise problematic, that group should be at the top of your list to contact.

1. Talk with colleagues who have dealt with similar issues or review records of public hearings about related concerns for ideas about interested audiences.

2. Determine which audiences are most important for you to communicate with. It may help to prioritize your audiences by dividing them into three categories:

    a. The inner circle — those most likely to be very concerned and very interested. They must be contacted and involved to the greatest extent possible.

    b. The middle circle — those who have less concern or are more peripheral, but are apt to be upset if not contacted. They should be contacted, invited to be involved, and kept informed.

    c. The outer circle — those who are less likely to be concerned. This often includes the "general public." Less effort should be directed to these audiences than the other two, and the effort should be aimed at involving them in the middle circle.

    Note that the boundaries between these categories are not rigid; people who become more or less interested should be free to switch to a different circle. The categories are determined not by you, but rather by the community.

---

* Adapted from Chess, C., B. J. Hance, and P. M. Sandman, *Planning Dialogue with Communities: A Risk Communication Workbook* (Environmental Communication Research Program, June 1989).

3. As you contact people, ask them if they know of others you should be contacting.

4. In any mailing, public presentation, or media contact, indicate how interested people can get more involved.

5. Have sign-up forms at community meetings, leave some copies of the form in relevant offices, etc. And be sure to get back in touch with those who sign.

6. Put notices in newsletters of relevant organizations.

7. Put notices in your own employee and community newsletters.

⤴ Questions to Help Identify Key Audiences*

Audiences can be identified in various ways: by talking to colleagues and employees, by going through newspaper clippings and records of public hearings about related issues, through networking, etc. Some of the questions that might help you identify and prioritize your audiences follow. *Remember: Often the audiences that are most difficult to deal with — and the ones you might be hoping to avoid — are the ones you most need to communicate with.*

1. Which groups have been previously involved in this issue?
2. Which groups are likely to be affected directly or to think they are affected directly by the company's action?
3. Which groups are likely to be angry if they are not consulted or alerted to the issue?
4. Which groups would be helpful for you to consult with because they might have important information, ideas, or opinions?
5. Which groups should you involve to ensure that the company has communicated with a balanced range of opinion on the issue?
6. Which groups will others seek out for their opinions on the company's action (you don't want them blindsided or ill-informed)?
7. Which groups have responsibilities relevant to the company's action (e.g., firefighters, regulators)?
8. Which groups may not especially want input, but do need to know what your company is doing?

---

* Adapted from Chess, C., B. J. Hance, and P. M. Sandman, *Planning Dialogue with Communities: A Risk Communication Workbook* (Environmental Communication Research Program, June 1989).

✔ Potential Audiences*

The following list is meant to trigger your thinking rather than to be exhaustive. There may be other audiences in your particular situation.

1. **Government** — Federal, state, county, and municipal agencies and elected officials; legislative committees; quasi-governmental agencies such as sewerage authorities, regional planning commissions, and environmental commissions; emergency responders such as police and fire fighters.

2. **Employees** — And their families. Also retirees.

3. **Geographical Neighbors** — Local residents and businesses.

4. **Environmental Groups** — National, state-wide, and local groups; specific issue groups such as Superfund or siting groups; conservation groups dealing with watersheds, hiking, fishing, and natural features; groups with specific functions, such as legal, lobbying, research, or organizing.

5. **Civic Organizations** — League of Women Voters; associations such as Kiwanis, Rotary, etc.; associations of senior citizens; ethnic groups.

6. **Professional and Trade Associations** — Health professionals (doctors, nurses); technical people (sanitarians, water purveyors, consultants, planners); business organizations (realtors, chambers of commerce, industrial and agricultural groups).

7. **Educational and Academic Organizations** — Colleges; agricultural extension; public and private schools; academic experts in fields relevant to the company's action.

8. **Community Organizations** — Social service groups, etc.

9. **Religious Organizations.**

---

* Adapted from Chess, C., B. J. Hance, and P. M. Sandman, *Planning Dialogue with Communities: A Risk Communication Workbook* (Environmental Communication Research Program, June 1989).

◆ WHEN COMMUNICATING IN EMOTIONALLY CHARGED SITUATIONS, PUT PARTICULAR EMPHASIS ON LISTENING AND RESPONDING PERSONALLY

Dealing with emotionally charged situations is often a matter of personal style. However, attitude is critical. Most practitioners who have faced angry and emotional crowds say that what makes the difference is a genuine desire to listen to people, to relate as a human being rather than as a representative of a company, and to be of some help to citizens. There is general agreement that if people are upset — even though the company might not agree with them — they deserve to be heard and responded to. As Martin Ferris of Air Products and Chemicals says, "I've spent hours in living rooms with parents, listening to concerns...."

Ben Woodhouse of Dow Chemical says he used to feel that responding personally and admitting the validity of people's concerns would put his company in a position of liability. "I used to go into situations with the attitude that there was no reason at all for concern," says Woodhouse, who acknowledges that this way of responding damaged his relationships with the community. Susan Dupree, also of Dow, agrees that there is a risk in showing you have heard people's emotions and concerns, but that taking that risk is important for open and honest communication. Dupree adds that at Dow "we have permission to be risk-takers" in interactions with the community.

Keith Fulton of Exxon says that as a manager he feels he sets the tone for the people he takes with him when he goes to a meeting with citizens. If he goes in with a positive and willing-to-help attitude, says Fulton, the people he brings along are more likely to follow suit. On the other hand, if he were to transmit the attitude that citizens at the meeting will be unreasonable and therefore do not deserve to be heard, "somewhere along the way someone is going to blow a fuse." Most important, says Fulton, is recognizing that there will be a mix of people who are genuinely concerned and others who will be there simply to "blast the management of a big plant that's full of well-paid people. You're there to demonstrate...that you're willing to bend over backwards to be good citizens.... The vast majority of people will see that."

✔ How to Respond Personally

Responding personally is more than simply telling people that you know how they feel (do you really?), or that if you were in their situation you'd drink the water, breathe the air, or feel comfortable letting your child go back into the remediated school building. In fact, very often statements like these can backfire. Most practitioners say it is better to demonstrate that you care. Some suggestions from interviewees:

1. Know what people's concerns are and be prepared to address them.
2. Know how *you* feel and how you will react to certain issues.
3. As much as possible, deal with people in informal situations in which it is easier to act as a person rather than as part of the corporate "monolith."
4. Tell people who you are, what you do, and what you can tell them about the concerns they have voiced.
5. Try to visualize situations in which you had similar concerns for yourself, your family, or your property.
6. Be flexible; allow people to help you structure the agenda to fit their needs.
7. If someone is introducing you, make sure the person sets the stage for you to be perceived as a human being, not just the "expert" or the "company representative."
8. Keep your temper; resist the urge to respond to barbs defensively. Try to get at the concern or question behind the barb.
9. Try not to get *colder* when the questioning gets hotter or to retreat behind jargon and your professional stature.
10. Don't be afraid to show people that you understand where they are coming from if you do (and never, never say you do if you don't).
11. Acknowledge people's emotions explicitly — say that you can see that they are frightened or angry.
12. Acknowledge shared values. Before talking about whether a child's leukemia could have resulted from plant emissions, acknowledge the horror of a child being diagnosed with leukemia.

13. Answer people's questions without judging them.

14. Don't forget the human touches at meetings — food, child care, etc.

15. Where appropriate, explain how you personally would respond in this situation — but acknowledge that there are other reasonable responses.

16. Remember people's names and situations. Build a personal relationship with concerned citizens in the same way you would with colleagues you expected to see again.

17. Share in the life of the community. This doesn't replace performance, but it shows that you are a person too.

✔ Some Dos and Don'ts of Listening*

In a crisis situation, you will be faced with several different audiences requiring your attention and ability to really "hear" what they are saying.

Here are some suggestions for improving your listening skills, but be reminded that mastery of these skills requires repeated practice. When listening, try to *do* the following:

1. **Become aware of your own listening habits** — What are your strong points? What are your faults? Do you judge people too quickly? Do you interrupt too often? A better awareness of your listening habits is the first stage in changing them.

2. **Share responsibility for the communication** — Remember that it takes two to communicate — one to talk and one to listen — with each person alternating as the listener. Whenever you are unclear about what a speaker is saying, it is your responsibility to let the speaker know this, either by asking for clarification or actively reflecting what you heard and asking to be corrected.

3. **Be physically attentive** — Face the speaker. Maintain appropriate eye contact. Make certain your posture and gestures show you are listening. Sit or stand at a distance which puts you and the speaker at ease. Remember that the one who is speaking wants an attentive, animated listener, not a stone wall.

4. **Concentrate on what the speaker is saying** — Be alert for wandering thoughts. Being physically and verbally responsive will probably help you concentrate on what the speaker is saying.

5. **Listen for the total meaning, including feelings as well as information** — Remember that people communicate their attitudes and feelings "coded" in socially acceptable ways. Listen for the feelings as well as the content.

6. **Observe the speaker's non-verbal signals** — Watch the speaker's facial expressions and how much he or she gazes and makes eye contact with you. Listen to the speaker's tone of

---

* From Eastwood Atwater, *I Hear You* (Englewood Cliffs, NJ: Prentice Hall, 1981). ©1986 Eastwood Atwater.

voice and rate of speech. Does the speaker's body language reinforce or contradict the spoken words?

7. **Adopt an accepting attitude toward the speaker** — An accepting attitude on the listener's part creates a favorable atmosphere for communication. The more speakers feel accepted, the more they can let down their guard and express what they really want to say. Any negative attitude on the listener's part tends to make a speaker feel defensive, insecure, and more guarded in communication.

8. **Express empathetic understanding** — Use active, reflective listening skills to discover how other people feel and what they are really trying to say in terms of their own frame of reference.

9. **Listen to yourself** — When you recognize the feelings stimulated in you by another's message and can express those feelings, this clears the air and helps you to listen better.

10. **"Close the loop" of listening by taking appropriate action** — Remember that people often speak with the purpose of getting something tangible done — to obtain information, to change your opinion, to get you to do something. The acid test of listening is how well you respond to the speaker's message with an appropriate action. In listening, actions speak louder than words.

While emphasis should be on positive suggestions for improving listening habits, it is helpful to keep in mind some of the pitfalls of listening. Consequently, in listening, *don't* do the following:

1. **Don't mistake not talking for listening** — People who remain silent aren't necessarily listening. They may be preoccupied with their own thoughts. On the other hand, people can talk a lot and still process information and listen quite well.

2. **Don't fake listening** — Whenever you try to fake listening, your disinterest or boredom inevitably shows up in your facial expressions or body language. More often than not, fake listening comes across as an insult to the speaker.

3. **Don't interrupt needlessly** — People in positions of power tend to interrupt more often than those not in power without realizing it. If you must interrupt someone in a serious conver-

sation, try to follow with a retrieval — helping the speaker to reestablish the train of thought.

4. **Don't pass judgment too quickly** — Judgmental remarks invariably put others on the defensive, serving as barriers to effective communication.

5. **Don't make arguing an "ego-trip"** — Even if you argue only "mentally" with what the speaker is saying, you tend to stop listening and look forward to your turn to talk. When you begin to argue verbally, you become so preoccupied with justifying your own views that you often fail to hear the other's viewpoint. When you honestly disagree, you need to listen carefully in order to understand what you are disagreeing with. Then state your point of view.

6. **Don't ask too many questions** — Closed questions that require a definite answer should be kept to a minimum. Even open questions that encourage a speaker to elaborate on a point should be used with caution. Too many questions have a way of shifting control of the conversation to the listener, putting the speaker on the defensive.

7. **Don't ever tell a speaker "I know exactly how you feel"** — This remark serves more to justify your own efforts than to convince someone you are really listening. In the first place, it is difficult to know just how another person feels. Then too, such a generalized remark is likely to distract the speaker from further efforts at self-expression as well as cast doubt on your own credibility as a listener. It is usually more effective to demonstrate you have heard with a reflective, empathetic response such as "I sense that you are feeling disappointed," or "I get the impression you are angry about this."

8. **Don't overreact to emotional words** — Be careful not to let yourself get so caught up in the speaker's outburst of feelings that you miss the content of his or her message. Be alert for loaded words and expressions; listen also for the message that comes with them. Your own feelings can block your understanding of something you may really need to hear.

9. **Don't give advice unless it is requested** — Even when someone asks your advice, it is better to use reflective listening skills to determine what that person wants to know.

10. **Don't use listening as a way of hiding yourself** — People may use the appearance of listening as a way of avoiding emotional involvement and real communication. The "listener" who uses silence as a personal retreat is inadvertently preventing effective communication, rather than furthering it.

♦ **IF IN DOUBT ABOUT WHEN TO RELEASE INFORMATION, LEAN TOWARD EARLY RELEASE**

Many industry practitioners believe that there are sometimes good reasons to hold onto information rather than to release it. The two most common reasons are (1) fear of getting the public stirred up for no reason, and (2) fear of going forward with incomplete or potentially inaccurate information. When to release information is situational. However, companies rarely make the mistake of releasing information too soon, while nearly all industry representatives are veterans of at least one battle that resulted from holding onto information longer than they should have.

A typical example of the kind of mistrust and outrage that is generated by a company's withholding of information comes from Greg Schirm of the Delaware Valley Toxics Coalition. Schirm tells of a water company that didn't inform customers that it was in violation of new, stricter water standards. In fact, the information came out inadvertently in a newspaper article. Not only was the company required by federal law to reveal this information, but customers felt it had an obligation to inform them of the elevated levels of contaminants *before* the new standards went into effect. Says Schirm, "The big issue...was 'Since...you've known that these [federal and state] standards were going to go into effect by a certain date...why did you wait until they went into effect....?' What people quickly realized was that...the company knew about them two years ahead of time."

Often companies feel they want to have complete information before they go public with it. Kelli Kukura of Du Pont says that it is critical to have enough information to answer people's questions before you release. When Du Pont called a public meeting to release information about its proposed on-site incinerator, says Kukura, the company "didn't have all the information they needed to go out.... Now we realize we should have provided more information through facts sheets, through the newspaper...." But while having enough information is important, companies sometimes use insufficient information as an excuse for withholding what they do know. It is generally an advantage to release provisional information early and ask for input, rather than to wait to release information that people are already demanding or that they will later resent you for having withheld.

✔ Ten Reasons to Release Information Early*

Decisions about when to release information depend in large part on the situation. However, companies should seriously examine the implications of holding onto information. The next time you contemplate whether to make information public, consider some of the reasons to release information early:

1. People are entitled to information that affects their lives.
2. Early release of information sets the pace for resolution of the problem.
3. If you wait, the story may leak anyway. When it does, you are apt to lose trust and credibility.
4. You can better control the accuracy of information if you are the first to present it.
5. There is more likely to be time for meaningful public involvement in decision-making if the information is released promptly.
6. Prompt release of information about one situation may prevent similar situations elsewhere.
7. Less work is required to release information early than to respond to inquiries, attacks, etc. that might result from delayed release.
8. You are more apt to earn public trust if you release information promptly.
9. If you wait, people may feel angry and resentful about not learning of the information earlier.
10. People are more likely to overestimate the risk if you hold onto information.

* Adapted from Chess, C., B. J. Hance, and P. M. Sandman, *Improving Dialogue with Communities: A Short Guide for Government Risk Communication* (Trenton, NJ: Department of Environmental Protection, January 1988).

✔ How to Release Information Early

Often the question of when to go to people with information has more to do with sequence (who to go to when and in what order) than with actually withholding information completely. Many companies are willing to go to a state agency with a plan, or to local officials with information, thinking that this constitutes releasing information. What often happens, however, is that the information is leaked to the press, and communities then perceive — often correctly — that decisions are being made without their input, which they resent.

1. Leaning toward early release, tailor the sequence to fit the situation.
2. Try to release while answers are still provisional so people feel you have thought about options but there is still an opportunity for input and change.
3. Anticipate questions and make sure you have answers or are able to detail what you are doing to get answers.
4. Pull together a group of managers and go through a scenario of how the data will come out, then make sure that is the way you want it to happen.
5. Don't blindside anyone who is involved or may have a particular interest. Local officials, environmentalists, and employees, for example, should hear from you rather than the media.
6. To avoid rumors, release information to all audiences, including the media (see pages 109 to 112) within several hours if possible, even if it takes a team of people to go out to individual audiences.

✔ What To Do When You Can't Release
Information Early

1. Release what you do have before you are asked.
2. Explain why you do not have or cannot release all the information.
3. Commit to a schedule for release of information — and keep to it.
4. Talk about what actions you are taking. Ultimately, you will be judged by what you do, not only by what data you release.
5. Tell people what precautions they can take while waiting for more data.

◆ FIND INFORMAL AS WELL AS FORMAL WAYS TO INTERACT WITH PEOPLE

The number and types of ways for companies to get involved in talking with the community are virtually unlimited and require a less sophisticated structure than many might think. Often it is simply a matter of brainstorming how a company can get out and talk with as many groups in as many different ways as possible. Many companies are seizing any opportunity to interact with the communities in which they reside, from one-on-one telephone calls and home visits to more elaborately planned "neighborhood nights" and open houses. The success of any of these forums depends on a number of factors, but probably the most important is the willingness of the company not just to present information, but also to be open to listening and responding to people's concerns.

Plant tours and open houses, long-accepted public relations activities, are now being used by companies to listen to the public. An example of a more formally planned interaction is an open house held by General Electric Silicones in Waterford, Connecticut. According to H.F. (Bud) Lindner, the company held an "environmental open house," inviting the public through newspaper advertisements and sending out invitations to town residents. Over 3500 people attended the open house, which included tours of the waste treatment facility and was staffed by plant employees, who could answer people's specific questions. While plant tours alone cannot substitute for smaller and less formal interactions with the community, they can help open up the lines of communication and help the company get a better sense of the community for future interactions.

More informally, Du Pont has begun sponsoring neighborhood "coffee talks," held in the evenings for 12 to 15 local residents to discuss Du Pont's proposed incinerator for the Deepwater plant. Kelli Kukura notes that the public meeting forum Du Pont used when it first announced the incinerator was not appropriate for providing information and feels smaller groups would have been more appropriate at that time. These informal gatherings, according to Kukura, are held for interested neighbors by plant employees, retirees, etc. in their homes. The plant manager and the incinerator project manager attend to answer the questions and concerns of people who may not be associated with any of the community groups to which the plant normally sends speakers.

Some innovative ideas have been tried at the Rohm and Haas facility in Bristol, Pennsylvania. In July 1986, the plant opened an information center at the front gate which contains all environmental studies conducted by the company, technical journals, government reports, videotapes, etc. There is a telephone people can use to call for more information. The company has also established an advisory committee of local residents which meets regularly to discuss "a whole range of issues about how we operate," says Brian McPeak, manager of public affairs.

In order to bring outside points of view into the company, Dow Chemical's issue management team, with representatives from the environmental, legal, manufacturing, and public relations areas of the company, meets every other month with outside speakers such as legislators, environmentalists, and academic experts. Says Ben Woodhouse, "It's designed to make sure we have our house in order on the issue, and...to go out and shape the issue outside the company. We want to be an active player in the policy process." By using several different forums to provide and receive new information, Woodhouse says, the company is "trying to move from being part of the problem to being included and considered part of the solution."

✔ Forums for Dialogue

In order to communicate effectively, companies must develop ways to listen, as well as to deliver their messages. Below are some ways to build two-way communication into traditional outreach efforts — as well as a few less obvious approaches.

1. **Telephone calls** — Often picking up the phone and asking a few questions, particularly when you are pressed for time, makes more sense than waiting for a meeting or communicating more formally.

2. **Open houses** — Rather than "dog and pony" shows, these can be a way of listening to community concerns, recording them for later consideration, and responding to them.

3. **Neighborhood nights, coffee klatches, etc.** — These informal gatherings can be a great way to find out community concerns and to begin a dialogue.

4. **"Breakfast meetings" with local officials** — Du Pont uses these informal early morning meetings to update local officials, but also to keep apprised of their concerns.

5. **Plant tours** — Tours can encourage people to ask questions and raise issues. Some plants have conducted tours for community groups accompanied by their own technical advisor.

6. **Speakers bureaus** — Company employees can listen as well as "speak." They should be sure to communicate internally what they learn.

7. **Employee and community newsletters** — These can solicit questions and topics for future issues.

8. **Physicians' panels** — Exxon Baytown and other companies have brought together local physicians to talk about their concerns.

9. **Citizen advisory committees** — Care must be taken to ensure that boards represent all interested groups and are a way of soliciting advice, rather than promoting your agenda (see page 31).

10. **Call-in radio and television programs** — These can be an opportunity to get a feel for the range of opinion in the community.

11. **Teachers' seminars** — These can be used to solicit and discuss local environmental and industrial issues relevant to teachers and their students.

✔ Preparing for a Meeting or Other Community Gathering*

The major areas for concern when preparing for any type of community gathering — whether it is an informal meeting or an open house — include process, content, logistics, and trouble-shooting. Attention to all of these areas is important; neglecting to think about any one of them may lead to a less-than-favorable outcome. For example, if you have failed to provide parking or if you have neglected to invite an interested and affected group, people may be angry at you even before you give your well-prepared presentation.

Clearly, the planning for an open house or a plant tour will be more extensive than for a coffee klatch. However, the following lists will give you items to consider when planning any of these forums.

Process

1.   Talk with affected people ahead of time.
2.   Do appropriate outreach to see that those who should be there — and who want to be there — are invited.
3.   Arrange for appropriate spokespeople from your company.
4.   Think about the possibility of co-chairing the meeting with someone from the community, an environmental group representative, or perhaps a neutral party.
5.   Pick a suitable location (a neutral location may be more appropriate than somebody's "turf").
6.   Develop an agenda that provides time for audience concerns and builds in flexibility to deal with unforeseen topics that might arise. Better yet, develop the agenda with representatives of the audience.
7.   Make sure people have had some vehicle to express their concerns before you make a long formal presentation — particularly if tension and mistrust are high — either before the meeting or at the opening of the meeting.

* Adapted from Chess, C., B. J. Hance, and P. M. Sandman, *Planning Dialogue with Communities: A Risk Communication Workbook* (Environmental Communication Research Program, June 1989).

8. Get input from your audience(s) about the agenda before the meeting and revise as needed.

9. Consider how you will handle conflict if it arises.

10. Appoint a notetaker (someone should write down promises company representatives make and follow up on them).

11. Consider how you will get feedback on the effectiveness of the gathering, e.g. evaluation forms, etc.

12. Develop a strategy for dealing with the media so that you can respond to their needs and avoid disrupting the gathering.

Content

1. Make sure you know what issues your *audience* wants addressed, not just those *you* think are relevant.

2. Go over a list of possible questions in advance and be prepared to respond to them. Better yet, integrate the answers into your presentation, but still allow plenty of time for questions.

3. Prepare background material and handouts.

4. Get feedback on your presentation from someone not involved with the issue.

5. Evaluate the materials you have developed.

6. Know whether the company is in compliance with existing regulations and mandates, and, if not, be prepared to outline what you are doing to achieve compliance. Be prepared to address other substantive concerns (see pages 80 and 81).

7. Find out what the company has promised at past meetings — especially questions you promised to get answered — and make sure your presentation responds to those commitments.

Logistics

1. *Location* — Should be convenient with sufficient parking. Make directions available. Building and room should be wheelchair accessible. Room should be large enough and temperature controlled.

2. *Support* — Enough and right type of company representatives should be present. Buses should be available to take people

around the plant if there is a tour. All necessary safety precautions should be taken.

3.   *Equipment* — Audio-visual equipment should be tested, and spare parts should be readily available. Microphones should be set up for speakers and audience. Podium, tables, chairs, displays should be set up ahead of time.

4.   *Time* — An appropriate time should be arranged, making considerations (i.e., evenings or weekends) for working people.

5.   *Food and beverage* — Should be available.

6.   *Other* — Make childcare available, prepare name tags for speakers, prepare a sign-in sheet.

Trouble-Shooting

You also might want to do some thinking *before* the meeting about how you will react and respond *during* it. Some things you may want to be prepared for are:

1.   Going over the agenda at the beginning of the meeting and, to the extent possible, making changes that people suggest.

2.   Changing gears in your presentation based on audience reaction.

3.   Dealing with outside groups you hadn't invited or counted on.

4.   Being prepared to respond to suggestions, concerns, and requests.

5.   Being prepared to stay after the meeting to answer individual questions.

6.   Handling conflict if it erupts.

7.   Dealing with *more* people than you expected to attend.

8.   Dealing with *fewer* people than you expected to attend.

9.   Dealing with someone in the audience who starts giving a speech.

10. Dealing with people who monopolize the meeting.

11. Dealing with the media.

◆ DEVELOP AN OUTREACH PROGRAM RATHER THAN WAITING TO RESPOND

Many companies are doing some type of outreach, and some are fairly advanced in this respect (see page 27). Most companies have been doing outreach as part of the chemical industry's Community Awareness and Emergency Response (CAER) program, while others have used the Title III reporting requirements to be proactive in their communications. While some companies argue that communities aren't interested except when there is an event or a crisis, most, according to Du Pont plant manager Richard Stewart, realize that "credibility must be established in times of non-crisis. In the past we've had a tendency to respond in a crisis when we've expected something from the community. What you learn is that communication is an ongoing process."

Martin Ferris of Air Products and Chemicals, Inc. says that his company used the release of its Title III information as an opportunity for outreach to the community. In May 1988, the company sent out 50 employees to hang information packets on the doorknobs of all houses within a $2^1/_2$-mile radius of the plant. The packet consisted of informational brochures on the plant and SARA Title III, a letter from Ferris inviting residents to come to the plant for a tour, and a response card. In addition, Air Products approached county officials, superintendents of schools, and local community and church groups. Although the company had a low, but generally favorable, response from the community, Ferris feels continued outreach is important because "people are entitled to understand what's around them. If you don't [do outreach], you'll be represented by somebody else in terms that may not be favorable to the industry or to the betterment of people's understanding...and it's the right thing to do." Also, people want to know they have access to the company, whether or not they take advantage of it. A low response may mean that people support the company's openness, and shouldn't be interpreted as a sign that people are apathetic and don't appreciate being kept informed.

Environmental groups agree that it is to the company's advantage to do outreach rather than waiting to respond. "I just assume that it's in the company's interest always to put their spin on [information] early," says Fred Millar of the Friends of the Earth/Environmental Policy

Institute/Oceanic Society, who also feels that companies can use Title III to do outreach in meaningful ways. "They ought to cooperate fully with local emergency planning committees (LEPCs) and lend all kinds of help, including computers and math and printing...." Millar says it is important for companies to recognize the value of the LEPCs rather than "starting side committees that are company-dominated and have resources that the LEPC does not."

At the time the Title III deadlines for releasing information were drawing near, CIBA-GEIGY decided to be proactive about releasing its Title III data in Toms River, New Jersey, even though the company was embroiled in controversy over its ocean discharge pipeline and the proposed construction of a pharmaceutical facility on its grounds. Thomas Chizmadia, director of corporate communication and public policy, views a proactive stance as necessary for getting the information out before the media have to dig for it, but also to honor the commitment the company has made to its neighbors. Although some feel CIBA-GEIGY's Title III communication would have had more credibility if the company had been perceived as effectively handling its other environmental problems, the company nonetheless chose to take the lead on this issue rather than retreat.

✔ Tips for Doing Outreach

Reaching out to a community can open a dialogue that is productive to the community and to the company. The following are some considerations in thinking about your outreach program.

1. **Speak with the doubters inside the company** — Talk to those who feel you really *should* let sleeping dogs lie and help them to understand the value of community outreach (see pages 127 and 128).

2. **Aim at many small events** — While open houses are useful, don't try to meet outreach needs with one big bash.

3. **Don't worry if response is small** — Simply inviting people in helps to build trust and confidence — even in those who decide not to come.

4. **Do as much outreach as possible on others' turf** — Offer to speak at other people's meetings, set up a booth at their fairs, etc.

5. **Make sure you address the tough issues** — Get the problems and conflicts out on the table and clear the air. Outreach that seems like an effort to paper over problems can do more harm than good.

6. **Be clear about your message** — Design your program, keeping in mind what you want to communicate, but also the messages the community may want to send or hear about.

7. **Give the program enough resources to make it work** — Don't just go through the motions to prove you have a program.

8. **Listen** — Your outreach program can be a way to learn.

Yes, but...

It makes no sense to seek out people who oppose us.

It is true that it is rarely easy to deal with people who distrust you and oppose your past actions or future plans; but you probably won't be able to avoid them. You can choose to fight it out at a distance — in headlines or public meetings — or try to understand each other across a table.

If you open a door to your company, opponents will feel less of a need to knock it down. There is more likelihood that you can discuss issues of substance rather than why your door was closed. Although some opposition groups may decline your invitation, that should be their decision. Your willingness to enter into a dialogue should be visible to the community. If you act reasonable, others — among your opposition, your allies, and the undecided — are more apt to see you as reasonable.

Dealing with people's emotions will validate their "wrong" assumptions.

There is a difference between listening (and showing you are listening) to someone's pain and accepting responsibility for that pain. People can care and continue to disagree.

There are often real issues — technical issues or values issues — behind people's concerns. But even if there is *no* real issue you can discern, just misunderstanding and fear, the best approach is to take their concerns seriously.

If we release information early, we are more apt to cause undue alarm.

You may cause greater alarm, compounded by resentment and hostility, if you hold onto information. In addition, when people are not given information, they may think the truth is too awful to be told. Instead, consider releasing information in context and with caveats, if necessary.

I'm technically trained and up to my ears in my "real" job. I can't possibly do all this.

It probably does feel unfair that many environmental managers and technical staff are now asked to take on tasks outside of their areas of expertise, tasks that they probably have relatively little training for. Although this may feel like changing rules in the middle of the game, it is increasingly clear that playing by the old rules is not going to work.

4   TALKING ABOUT
THE NUMBERS

♦ FIND OUT WHAT INFORMATION PEOPLE WANT AND WHAT CONCERNS THEY HAVE

Communication efforts often misfire because companies do not take the time to find out what people's concerns are and what information they want. Experts within the company often frame what they think are the key issues and set about preparing written materials and presentations to address these issues. More often than not, the company will frame issues based on technical concerns and expect that their information will satisfy people. However, they often find once they go out to a public meeting or other forum that the material they have carefully prepared is ignored — or even worse, angers people. Companies are often surprised, and unprepared, when they find out the public's real concerns.

Joan Ebzery of Clean Sites, Inc., says that at the Rose PCB clean-up site in Holden, Missouri, an unanticipated concern of townspeople was the presence of the clean-up crew in the town. Says Ebzery, "People told me they were worried about the workers from Rose eating in the local restaurants — particularly those with salad bars. What they were raising was a very valid concern about possible ingestion of PCBs carried from the site."

Thad Epps of Union Carbide says that when his company wanted to site an incinerator at a company plant in West Virginia, it conducted a telephone survey in the area "to get the lay of the land," asking how residents felt about Carbide and how they felt about waste in general. The survey was followed by a series of small meetings with employees, then with local politicians, education leaders, business leaders, and citizens (whom Carbide found through networking among these other groups). The meetings had a facilitator not affiliated with Union Carbide, who basically told the group, "You know why we are here. What kinds of questions and concerns do you have?" The plant manager then gave a presentation and attempted to answer as many questions as possible. Then the facilitator went through the original list of questions and asked if there were others. Epps characterizes these meetings, for the most part, as "nonconfrontational," and says that people were "obviously pleased that we promised to keep them informed and have additional meetings if necessary to get their feedback."

✔ Identifying People's Concerns*

Listed below are some ways of getting a "snapshot" of people's concerns. No one of the approaches below will give you the full picture. Better to use two or more of these, and if the information conflicts, keep investigating. The following methods are most appropriate for modest risk communication efforts; larger communication efforts require more resource-intensive, methodologically stringent approaches.

1. Review newspaper clippings to get an overview of the issue.
2. Discuss audience concerns with colleagues who have dealt with similar issues or situations.
3. Meet informally with community members interested in the issue to get a first-hand idea of both substantive concerns and people's feelings about those concerns.
4. Send a letter to potentially interested people and organizations asking them to send you a list of their questions and concerns.
5. Develop a survey (which you can give to people through a door-to-door effort, at meetings, or in mailings) that asks people about their questions and concerns. (Develop the survey with care, so that people feel they can articulate their concerns, not merely respond to yours.)
6. Brainstorm questions and concerns at the *beginning* of a meeting, or ask people to write their questions on index cards that you distribute and collect.
7. Brainstorm questions and concerns at the *end* of a meeting. Or hand out index cards at the end of the meeting in preparation for the next.
8. Consult advisory committees (see page 31).
9. Conduct a poll (but recognize that polls are more useful to get general information from many people than to obtain in-depth information about people's concerns).
10. Conduct a focus group to elicit attitudes, ideas, and feedback.

---

* Adapted from Chess, C., B. J. Hance, and P. M. Sandman, *Planning Dialogue with Communities: A Risk Communication Workbook* (Environmental Communication Research Program, June 1989). Many of these methods are similar to those used to evaluate communication efforts. See the list of "Quick and Easy Feedback Methods" in Chapter 6.

✓ Questions Audiences May Ask of You*

In general, the types of concerns people will have over an environmental issue will fall into four categories: (1) health and lifestyle concerns (How will this affect me or my family?); (2) data and information concerns (What is this stuff?); (3) process concerns (How am I being treated?); and (4) risk management concerns (What are you going to do about this?). All four types of questions may be represented in any one community.

The following checklist represents some common concerns you might expect to hear when you do the research suggested in the previous checklist. *We provide it to familiarize you with the types of community concerns you may face, not as a substitute for identifying community concerns.* We cannot overemphasize that each situation is different and each community has its own set of specific concerns. Indeed, each individual *within* a community has his or her own concerns.

Health and Lifestyle Concerns

1. What is the effect of this chemical on my health?
2. What levels of this chemical are safe?
3. Are my children at any special risk? (Concerns about children are often primary and quite specific about the implications of exposure and whether certain behaviors will increase their risk.)
4. We've lived here for 20 years. Are we more likely to get cancer than people who have only been here for five years?
5. What studies have you done to support the health claims you are making?
6. We are already at risk because of X. Will Y increase our risk?
7. How will this affect our quality of life — property values, the stigma of X attached to our community, etc.?
8. How will we be protected in the event of an accident?

---

* Adapted from Chess, C., B. J. Hance, and P. M. Sandman, *Planning Dialogue with Communities: A Risk Communication Workbook* (Environmental Communication Research Program, June 1989).

Data and Information Concerns

1. How sure are you?
2. What is the worst-case scenario?
3. What do these numbers mean and how did you get them?
4. How do we know your studies are correct?
5. What about other opinions on this issue?
6. How do our exposures compare to the standards?
7. You say *X* can't happen. Why not? And what will you do if it does?

Process Concerns

1. How will we be involved in decision-making?
2. How will you communicate with us or reach us in an emergency?
3. Why should we trust you?
4. How and when can we reach you?
5. Who else are you talking with?
6. When will we hear from you?

Risk Management Concerns

1. When will the problem be corrected?
2. Why did you let this happen and what are you going to do about it?
3. What are the other options? Why do you favor option *X*?
4. Why are you moving so slowly to correct the problem?
5. What government agencies are involved and in what roles?
6. What kind of involvement will we have?

◆ PUT THE RISK INTO PERSPECTIVE FOR THE COMMUNITY RATHER THAN TRYING TO MINIMIZE IT OR TO CONVINCE PEOPLE TO ACCEPT IT

Putting information into a meaningful perspective for a group requires that you first know something about the people to whom you will be talking — what their concerns are and what kind of information will help them understand and make decisions (see pages 73 to 77). Even more fundamentally, it requires you to accept the goal of *informing* people so they can make wiser judgments, rather than trying to preempt their judgments and impose your own. Communities often feel that technical information is presented in a way that is meant to make them feel that the risk is less serious than it may be. Emphasizing how small a number is by using comparisons that trivialize the risk does more to outrage people than to enlighten them (see page 91). Presenting information in a way that ignores people's concerns as well as their intelligence doesn't work.

Fenceline data are used by some practitioners to put information into perspective for the community. Many companies have air monitoring equipment at their property lines and compare the data they get from those stations to government standards. Some feel this is more effective than modeling because actual release levels are measured. However, Colin Park of the Dow Chemical Company points out that fenceline data alone may also be misleading if the monitoring stations are not located in the emission plume. Park suggests that combining fenceline data with modeling data may provide more meaningful information to the company and to the community. "Monitors are good to provide *assurance* to the public that we are attempting to measure emissions as well as calculate them [with modeling]."

More important, some companies also provide information on steps they are taking to alleviate exposure. GE Silicones, for example, tells people of its agreement with the local water company to alert it immediately when on-site monitoring data read over a specified number. In this way, the water company can switch to an alternative source until the problem is found and resolved. While finding ways to put the numbers into perspective is important, what really matters to people is what you are doing to reduce, prevent, or mitigate exposure.

Environmentalists resent companies' failure to provide as much information as possible on potential health risks, even on chemicals for which the company is in compliance. According to Greg Schirm of the Delaware Valley Toxics Coalition, simply assuring the community that the company is below the standard for a chemical "doesn't really [tell] people anything significant. You are telling them only that you want them to believe that they're safe. You are not giving them enough information to make much judgment about it."

Fred Millar of the Friends of the Earth/Environmental Policy Institute/Oceanic Society says that often communities ask for *hazard* information but are supplied only with the company's estimate of the *probability* of an accident happening. For example, says Millar, in Torrance, California, the Mobil Corporation issued a press statement which discussed the low probability of a release of hydrogen fluoride from its refinery, but failed to supply the community with hazard information (for example, plume maps and worst-case scenarios) that would describe the catastrophic consequences of such an incident. Millar says that it is impossible for the community to plan for emergencies without this critical information.

Ken Brown of the New Jersey Environmental Federation feels that people should have information on chemicals they may be exposed to regardless of whether the exposure is within government guidelines. Brown adds that often it is possible to "do better than the law" — that in areas in which companies have accomplished source and waste reductions, companies were willing to go the extra step. Also, adds Millar, because the government's credibility is low, "more informed citizens will not be mollified by a company's meeting the standard."

✔ Things to Explain about Risk Data

The following are some of the specific questions people may ask you about the risk. However, as the first guideline in this chapter points out, people may have concerns beyond the "data," and it is best to try to anticipate and address those concerns.

1. What the chemical is. Chemical composition, taste, smell, etc.
2. What it is used for, at the plant and elsewhere.
3. How much of the chemical is stored, escapes, and is emitted each year.
4. In what concentration it is found: in the air, in the water, in the ground, at the fenceline, at city hall, on a bad day, on a typical day, at the height of an accident.
5. What chronic and acute effects it can have.
6. What exposure routes are particularly problematic in this situation.
7. Who is especially at risk. For example, children, the elderly, fetuses, asthmatics, etc.
8. How much concentration for how long will cause what health problems.
9. How good the data are. Animal studies, epidemiological studies, monitoring stations, etc.
10. What is being done to collect better data in the future.

✔ Things to Explain Besides Risk Data

1. Whether the risk is higher or lower than in the past and whether it is likely to get higher or lower in the future.
2. What the company is doing to monitor the risk, to reduce the risk, and to prevent an accident from happening.
3. How the community will know these measures are being put into place.
4. What the timetable for completion of these measures is.
5. How the company will get information to the community.
6. Which people in the company members of the community can call to get information and to report any problems.

♦ USE CARE WHEN SIMPLIFYING RISK INFORMATION

Simplifying risk information is a necessary evil in getting complex technical concepts across to the lay public. However, practitioners' attempts to simplify technical information often go too far (by leaving out important information) or not far enough (by failing to recognize exactly what needs to be simplified). Knowing something about your audience before you prepare your presentation is extremely helpful, but at the very least you should have a clear idea of the major points you need to cover and try to state them simply and clearly, leaving out ancillary material that will not be helpful to people's understanding of the basics. (Of course, if people ask for more, you should be ready to give it to them.)

One way to simplify technical material is to simplify the language used to describe it. While most technical people will agree that eliminating jargon from a technical presentation is important, far fewer are able to do it. Even the conscientious often slip into using jargon when they are in formal situations, or out of defensiveness, or as a short-cut. Martha Bean of CH2M Hill says, "It's hard to recognize jargon when we're using it." She points out that using "six-syllable words and five-letter acronyms" is second nature to technical people and they forget that "it took them years of training to understand and remember" the terms they often expect community members to learn in one meeting. Bean provides the example of an effort to translate technical information into Spanish. When someone made the point that there were no Spanish equivalents for the English phrases "remedial action," "confined aquifer," and "SNARL" (Suggested No Adverse Response Level), Bean realized that "we couldn't depend on the comforts of jargon when working in another language." Then she realized that, in truth, there are no easily understood English equivalents of these terms either. Bean adds, tongue in cheek, "We should all go through the exercise of translating all our communication 'into Spanish,' [and whatever can't be translated] we shouldn't use in English either."*

The quality of the graphics you use will determine whether they

---

* Bean, M., "Speaking of Risk," presented at the California Manufacturers Association Risk Assessment and Risk Communication Conference, Los Angeles, CA, May 1989.

amplify or obscure your message. Many practitioners find graphics particularly useful in showing groundwater flows, routes of exposure, schematics of plant processes, etc. However, poorly designed and executed graphics do more to confuse than to enlighten. Graphics should be simple, and illustrate only one point per frame. They should be used to make a point or to clarify a point, not to provide self-serving documentation for one. Some common complaints about graphics are that they contain too much information on one slide or frame, that columns of numbers are too long and decimal points don't line up, and that slides can't be read from the back of the room. Poor graphics can make an audience feel frustrated and angry and seriously hamper your presentation.

✔ Simplifying Language

The following are some tips to help you cut the jargon out of your communication with people.

1. Ask yourself if you are using the word to impress. If so, cut it out.

2. Once you have explained the piece of jargon satisfactorily, consider dropping it and using the plain language equivalent instead. (Thus, don't simply define "potable water" as drinking water — use "drinking water" throughout.)

3. If the jargon is needed by the audience, either for convenience or efficiency or to understand printed materials on the subject, teach it. But only then.

4. If you have to teach the jargon, introduce the concept first in plain language, then say there is a technical term for it. Never use the term first before you explain what it means.

5. Be especially wary of jargon when the conflict is heating up or when other professionals are present. This is when technical people use jargon the most — and when citizens resent it the most.

6. Acknowledge to your audience that you may forget and use jargon. Ask your audience to stop you immediately if you lose them.

7. Work with other presenters to help them simplify similarly.

8. After you've gotten rid of the jargon, work on other complicated language, sentence structure, etc.

9. For a written presentation, check readability by asking a non-technical person to read to make sure it is clear and simple enough. (Or use a more technical check for readability. See "Quick and Easy Feedback Methods" in Chapter 6). For a spoken presentation, practice on a nontechnical person.

✔ Simplifying Graphics*

Let's suppose that your plant has a groundwater contamination problem. You have collected and analyzed samples and have compared the results to standards based on human health criteria. Example 1 is from your internal report. Your task is to make a slide from your table for a public meeting called to discuss the contamination.

The graphics that follow in Examples 1 to 4 show a progression from an unclear table of numbers to a more focused, purposeful table. Some experienced speakers may judge the first two tables full of elementary mistakes; but in fact, often tables are full of such mistakes. The graph in Example 5 (while not without its own problems) illustrates yet another way to show the information graphically. The key to simplifying your graphics, says Martha Bean, is to decide what the "main message" is (which in this case is "Is the water safe according to standards?" and "Will I be hurt?") and then make sure it "is not obscured by detail."

It is not uncommon for presenters of technical information to make a slide simply by photocopying a page of a report or other document. However, as Bean states, doing that only obscures the information rather than highlighting it. Some of the problems with using Example 1 as a slide are (1) the numbers are too small; (2) there is too much information; (3) the table contains jargon; (4) the decimal points don't line up; and (5) the numbers are given in two different units.

Example 2 is not much of an improvement over Example 1 in readability, but at least it limits the data discussed from all parameters to only phenols and metals.

Example 3 is a substantial improvement in readability over Examples 1 and 2 because it uses a bolder, larger typeface and lines up the decimal points. It also uses only one unit (µg/l) and eliminates the surface water category, which isn't at issue here. On the other hand, Bean points out that this table still requires that people do their own comparisons to discover which chemicals are above the standard.

Example 4 is basically the same as Example 3, but the jargon is gone and the main message is spelled out both in the title and in the last column.

---

* Adapted from Bean, M., "Speaking of Risk," presented at the California Manufacturers Association Risk Assessment and Risk Communication Conference, Los Angeles, CA, May 1989.

EXAMPLE 1
Summary of Maximum Contaminant Concentrations for All Contaminants and Human Health Criteria (All Units in µg/l or ppb)

Parameter	Human health criterion	Highest observed levels	
		Ground water	Surface water
Conventional analysis			
Phenols, total	3,500	15,000	23
Metals			
Beryllium	0.037	15	—
Cadmium	10	770	—
Chromium	50	44	13
Mercury	0.144	0.4	—
Nickel	13.4	18	—
Lead	50	46	—
Thallium	13	93	—
GC/MS volatiles			
Benzene	6.6	10,000	9
Chlorobenzene	488	390	3
Chloroform	1.9	290	2
1,2-Dichloroethane	9.4	2,200	91
T-1,3-Dichloropropene 87	9	—	—
Ethylbenzene	1,400	1,700	—
1,12,2-Tetrachloroethane	1.7	5	—
1,1,2-Trichloroethane	6.0	390	4
Trichloroethene	23	44	—
Toluene	14,300	4,300	2
Vinyl chloride	20	400	5
GC/MS base/neutral			
Acenaphthene	20.0	68.0	5
Acenaphthylene	0.028	—	2
Anthracene	0.028	—	2
Benzo(A)anthracene	0.028	—	3
Benzo(B)fluoranthene	0.028	—	3
Benzo(A)pyrene	0.028	—	2
Bis (2-ethylhexyl) phthalate	15,000	190	37
Chrysene	0.028	—	2
1,4-Dichlorobenzene	400	6	—
1,2-Dichlorobenzene	400	6	—
Di-N-butylphthalate	34,000	—	2
Fluoranthene	42	6	2
Naphthalene	—	5900	—
Phenanthene	0.028	—	290
Pyrene	0.028	—	190
PCBs	0.00079	—	153

Adapted from Bean, M., "Speaking of Risk," presented at the California Manufacturers Association Risk Assessment and Risk Communication Conference, Los Angeles, CA, May 1989.

EXAMPLE 2
Summary of Maximum Contaminant Concentations for Phenols and Metals and Human Health Criteria (All Units in µg/l or ppb)

| | | Highest observed levels | |
Parameter	Human health criterion	Ground water	Surface water
Conventional Analysis			
Phenols, Total	3,500	15,000	23
Metals			
Beryllium	0.037	5	—
Cadmium	10	770	—
Chromium	50	44	13
Mercury	0.144	0.4	—
Nickel	13.4	18	—
Lead	50	46	—
Thallium	13	93	—

Adapted from Bean, M., "Speaking of Risk," presented at the California Manufacturers Association Risk Assessment and Risk Communication Conference, Los Angeles, CA, May 1989.

EXAMPLE 3
PHENOLS AND METALS:
Summary of Maximum Contaminant Concentrations and Human Health Criteria (All Units in µg/l)

CHEMICAL	MAXIMUM SAFE CONCENTRATION	FOUND IN GROUND WATER
Conventional analysis		
Phenols, total	3,500	15,000
Metals		
Beryllium	0.037	15
Cadmium	10	770
Chromium	50	44
Mercury	0.144	0.4
Nickel	13.4	18
Lead	50	46
Thallium	13	93

Adapted from Bean, M., "Speaking of Risk," presented at the California Manufacturers Association Risk Assessment and Risk Communication Conference, Los Angeles, CA, May 1989.

EXAMPLE 4
PHENOLS AND METALS:
Federal Safety Standards vs. What We Found (All Units in µg/l)

CHEMICAL	MAXIMUM SAFE CONCENTRATION	FOUND IN GROUND WATER	OVER THE MAXIMUM?
Phenols, Total	3,500	15,000	Yes
Metals			
Beryllium	0.037	15	Yes
Cadmium	10	770	Yes
Chromium	50	44	No
Mercury	0.144	0.4	Yes
Nickel	13.4	18	Yes
Lead	50	46	No
Thallium	13	93	Yes

Adapted from Bean, M., "Speaking of Risk," presented at the California Manufacturers Association Risk Assessment and Risk Communication Conference, Los Angeles, CA, May 1989.

Which Contaminants Will Require Clean-Up?

EXAMPLE 5. Which contaminants will require clean-up? Adapted from Bean, M., "Speaking of Risk," presented at the California Manufacturers Association Risk Assessment and Risk Communication Conference, Los Angeles, CA, May 1989.

Example 5 puts the same information into a graph, eliminating numbers and presenting a base point: the maximum level allowed. It removes the need for people to deal with numbers and gets right to a question they are likely to have: Which contaminants will require clean-up? This figure also focuses on *how much* above the standard each contaminant was.

On the other hand, points out Bean, there are still two potential problems with this graph. First, the scale is logarithmic and requires some explanation from the presenter. The stress on multiples of the standard could also be misleading. The figure shows that beryllium is 400 times the standard and cadmium only 77 times the standard. Does this mean beryllium is more of a risk than cadmium? Not necessarily, but this graph requires that the presenter address the point.

Bean adds that some practitioners have suggested that the final figure in this progression should be a simple list of metals that need to be cleaned up. In fact, such a list, says Bean, might well be presented *first* with supporting tables and figures available to provide more information once the bottom-line message is given.

◆ BE CAREFUL WHEN USING COMPARISONS

Risk probability numbers, especially tiny ones like 10^{-6}, are very hard for the nonexpert to understand and put into context. To help make these numbers meaningful, many industry spokespeople rely heavily on risk comparisons. Companies often rely *too* heavily on risk comparisons, in fact, mistakenly assuming that putting risk numbers into context is more central than explaining how the risk is being managed, dealing with community concerns, and other aspects of risk communication.

Comparisons to more familiar risks really can help clarify the data. Yet some of the most experienced risk communicators recommend extreme caution. "The only time I use them," says Keith Fulton of Exxon, "is when I'm giving speeches to the chemical industry and telling them not to use them." The problem in a nutshell is that industry risk comparisons too often strike citizens as belittling their concerns or as trying to coerce their acceptance of the risk. The classic example is the speaker at an angry public hearing who chides the audience for worrying about industrial effluents when they took a far greater risk driving to the meeting or smoking a cigarette during the break.

Whether a particular risk comparison works or backfires is largely a matter of context — the attitude of the company spokesperson, the suspicion of the community, the trust that has or hasn't been built, etc. Colin Park is a biostatistician and issues manager at Dow Chemical. In talking about animal toxicity data, Park often uses comparisons to make clear that animal research doses are much higher than human environmental exposures. Even a comparison that most communication experts would warn against, "like the 800 cans a day of saccharine," can be very useful in Park's judgment if it is handled carefully, "so you don't sound flippant or cavalier about it." Ben Woodhouse, Dow's manager of public issues, adds that it matters how much solid scientific information accompanies the potentially sensitive comparison: "It depends on what you wrap around it." But even Park and Woodhouse agree that it is unwise to tell a group of concerned citizens that the risk from industrial exposures is "safer than when you drove here to this meeting."

✔ Questions to Ask about a Risk Comparison

The experience of practical risk communicators suggests strongly that risk comparisons can help clarify risk statistics — and that comparisons can backfire badly. Unfortunately, there have been very few empirical studies to establish which sorts of comparisons work and which backfire under which circumstances. The following questions are no substitute for data, but they can help you reduce the percentage of backfires.

1. Does your comparison seem to be trying to preempt the decision about the acceptability of the risk? Are you acknowledging that risk acceptability is not a technical question, but a question of values, and that people have a right to participate in risk management decisions that affect their lives?

2. Are you really trying to make the size of the risk clear, or are you trying to "show up" opponents by making their concerns about the risk seem foolish?

3. If you were on the receiving end of this risk comparison — for a risk that concerned you and about which you knew very little — would you find the comparison useful, or would you find it irritating? Would you use this comparison with a nontechnical friend who was concerned?

4. Are you comparing a high-outrage risk with one that is low in outrage (see pagse 24 to 26) — for example, a coerced risk with a voluntary one, or a risk that benefits the people at risk with one that does them no good?

5. What is the relationship right now between you and your audience? Is it trusting enough that you can rely on the audience to see your comparison as an effort to clarify, or are they likely to misunderstand and take offense?

6. How sound is the comparison technically? Have you verified both halves — not only the data on the risk you are talking about, but also the data on the risk you are comparing it to?

7. Does your comparison rely on an unfamiliar risk of a familiar product or situation, such as aflatoxin in peanut butter? What will happen to your credibility when you try to explain that peanut butter is dangerous?

8.   How many points of similarity are there between the two risks being compared?
- Are they produced by the same industry or process?
- Do they lead to the same illness?
- Are they found in the same community?
- Are they solutions to the same problem?

The more related the two risks are, the better, especially when trust is low.

9.   Is your comparison "homey," snide, or slightly humorous — rolls of toilet paper stretching around the world, people getting struck by lightning in a snowstorm, etc.? If so, might some in your audience feel that you are talking down to them or belittling their concerns?

10.   Is your comparison likely to seem self-serving? If so, have you acknowledged that it is self-serving, that your company has a stake in convincing people to see the risk as you see it?

11.   What are you comparing? "Quantity" comparisons are usually helpful and inoffensive, while comparisons of concentrations, probabilities, or risks are more touchy. Similarly, comparisons aimed at making a particular risk seem large are much less sensitive than comparisons designed to show that the risk is small.

12.   On balance, is this risk comparison likely to seem useful and credible to this audience? A comparison that you suspect in advance the audience may reject or resent is a bad comparison, even if it is accurate. Is your goal to be "right," or is it to communicate effectively?

13.   Do you really need a risk comparison at all? Might you be putting too much emphasis on explaining your risk data and too little on other aspects of the controversy?

✔ Hierarchy of Risk Comparisons

In 1988, the Chemical Manufacturers Association (CMA) published a handbook devoted largely to risk comparisons.* The handbook featured a hierarchy of 14 kinds of risk comparisons arranged into 5 ranks from most to least acceptable. The rankings were based not on empirical research, but on the combined experience and intuition of ten risk communication scholars and chemical industry practitioners. One later empirical study, in fact, tested the recommendations in the handbook by asking people to select the best comparisons for a hypothetical plant manager to use and came up with different recommendations from the CMA's.**

Obviously the CMA's "hierarchy" of risk comparisons, summarized below, should be regarded as tentative; more research is needed on which risk comparisons work best in which situations. Obviously the answer does depend on the situation, and so the hierarchy of comparisons cannot replace the situation-specific questions on the previous page. Finally, the reader should not forget that risk comparisons are not the only way to explain risk data and that explaining risk data is not the only task in risk communication.

The hierarchy does provide a handy list of the types of risk comparisons you might want to consider using. But should you rely on the rankings? Some of the low-ranked comparisons are actually best at clarifying the risk, at least when both speaker and audience see the communication as an effort to clarify, not to coerce or belittle. In situations where tension and hostility run high, on the other hand, sticking to the high-ranked comparisons is probably wise, and attending more to the root causes of the tension and hostility is wiser still.

---

* Covello, V. T., P. M. Sandman, and P. Slovic, *Risk Communication, Risk Statistics, and Risk Comparisons: A Manual for Plant Managers* (Washington, DC: Chemical Manufacturers Association, 1988).

**Roth, E., G. Morgan, B. Fischhoff, L. Lave, and A. Bostrom, "What Do We Know about Making Risk Comparisons?" *Risk Analysis* (under consideration).

First-Rank Comparisons

1. Comparisons of the same risk at two different times. ("Forty percent less than before we installed the scrubbers last October....")
2. Comparisons with a standard. ("Ten percent of what is permitted under the EPA standard....")
3. Comparisons with different estimates of the same risk. ("Our best estimate of the risk is X, whereas the government's is Y and the Sierra Club's is Z.")

Second-Rank Comparisons

4. Comparisons of the risk of doing something vs not doing it. ("If we buy the newest and most advanced emission control equipment, the risk will be X, whereas if we don't buy it, the risk will be Y.")
5. Comparisons of alternative solutions to the same problem. ("The risk associated with incinerating our waste is X. The risk associated with using a landfill is Y.")
6. Comparisons with the same risk as experienced in other places. ("Our air toxic X problem is only one fifth as serious as Denver's.")

Third-Rank Comparisons

7. Comparisons of average risk with peak risk at a particular time or location. ("The risk posed by emissions of air toxic X on an average day is one thousandth as great as the risk last Wednesday, when a valve malfunctioned.")
8. Comparisons of the risk from one source of a particular adverse effect with the risk from all sources of the same adverse effect. ("The risk of lung cancer posed by emissions of air toxic X is roughly three hundreds of one percent of our total lung cancer risk.")

Fourth-Rank Comparisons

9. Comparisons of risk with cost, or of one cost/risk ratio with another cost/risk ratio. ("Saving one life by controlling emissions of air toxic X would cost Y dollars, whereas saving a life by improving particulate control would cost only Z dollars.")

10. Comparisons of risk with benefit. ("The chemical product whose waste by-product is air toxic X is used by hospitals to sterilize surgical instruments and thus contributes to the saving of many lives.")

11. Comparisons of occupational risks with environmental risks. ("The community is exposed to far less air toxic X than our plant workers, and medical tests at the plant show no adverse health effects.")

12. Comparisons with other risks from the same source. ("Our problem with air toxic X is no more serious than our problem with air toxic Y, which the community has long found acceptable.")

13. Comparisons with other specific causes of the same disease or injury. ("Air toxic X produces far less lung cancer than exposure to natural background levels of geological radon.")

Fifth-Rank Comparisons

14. Comparisons of two or more completely unrelated risks, especially if they disregard "outrage factors" like voluntariness, control, dread, and familiarity. ("The risk of emissions of air toxic X is far less than the risk of driving or smoking.")

◆ ACKNOWLEDGE UNCERTAINTY

The public's response to uncertainty often depends more on trust and credibility than it does on how uncertain the data are. Most practitioners say that dealing with uncertainty, while often problematic, is not as difficult when the community (1) feels every effort has been made to explain the area of uncertainty, (2) feels the company is doing everything it can to resolve the uncertainty as quickly as possible, and (3) feels steps are being taken to protect the public until more complete information is available.

Most practitioners feel that the way to deal with uncertainty is to admit that you are uncertain rather than to try to appear to know more than you do. Keith Fulton of Exxon Chemical says, "You can't be sure. You tell them that." For example, when Fulton was asked whether benzene emissions from the Exxon Chemical facility, which were below the standard, could cause cancer in the community, he replied, "Yes, it's possible. We don't think it's likely, but we don't know."

If there is uncertain information, it is far better to raise the problem yourself rather than waiting to be confronted with it. Too often, particularly in situations of low trust, companies seem to be hiding information if they fail to acknowledge uncertainty up front.

It is also important to be able to tell communities what you are doing about the risk — and what you are doing to get more information. According to Greg Schirm of the Delaware Valley Toxics Coalition, "Companies...need to talk about what it is they're doing, especially when there is an uncertainty. I don't think it's enough for a company to say 'we're not sure what the risk from this kind of release is or this chemical we are handling'.... The company needs to say 'we're not sure...so here's what we're doing...to minimize any risk from it.'"

A distinction needs to be made, however, between areas of scientific uncertainty and uncertain responses because the company has failed to do its homework. For example, many companies failed to collect emissions data before SARA Title III required them to do so because they were in a better position legally if they didn't know. In these cases, pleading uncertainty probably will not work with communities. As H.F. (Bud) Lindner of General Electric Silicones says, "It's O.K. to say you don't know, but it's better to know."

✔ Dealing with Uncertainty

1. Don't wait to be confronted. Acknowledge uncertainties up front.
2. Put bounds on uncertainty. Say the level is between two known quantities. Give error bars and confidence limits — for example, that you are 95% sure the correct estimate is between X and Y. Provide other people's estimates.
3. Make it clear that not all data are equally uncertain. Saying you are uncertain isn't the same as claiming ignorance.
4. Say what *is* certain. Tell what you do know.
5. Say what has been done to reduce uncertainty. Talk about how you have resolved related uncertainties and how that has helped the process. Discuss research you have already done and the ways it has reduced uncertainty.
6. Say what you will do to reduce uncertainty further. And say when the results of studies, new information, etc. will be available.
7. If the remaining uncertainty is very small or very difficult to reduce further, say so. Science is never absolutely certain, but if some findings (good ones or bad ones) are very nearly certain, say so, and explain that this is as close as the science comes.
8. Explain your cautiousness — and don't call your estimates or actions "conservativeness," a term people find confusing. Be overcautious until you are more sure, and talk about the margins of safety you are folding into your risk estimates to make sure that uncertainty doesn't lead to a safety problem.
9. Don't hide behind uncertainty. If it is more likely than not that the problem is real, say so.
10. Acknowledge and apologize if the company has dragged its feet. If you *should* have found out by now, admit it, and get cracking.
11. *Never* say, "There's no evidence of X" if you haven't done the study that tests the possibility.

Yes, but...

If we don't do our best to explain the risk so it seems as small as possible, the public will *really* blow the data out of proportion.

If you do your best to minimize risk, people will sense it and dismiss your explanations as "the industry line." In addition, they are bound to hear or read of other perspectives on the risk. If they don't hear these perspectives from you, they will assume that you withheld information to mislead them. In effect, if you do try to minimize the risk, you are less likely to convince many people and more likely to undermine your credibility. You are usually better off giving people more complete data with appropriate caveats, if necessary.

If the risk is actually smaller than driving a car or eating peanut butter, people should know it.

Comparing chemical or industrial risks to familiar ones can be useful to people. But when the comparison is used in a way that seems self-serving, it is likely to backfire. Unless your audience trusts you, you are better off leaving such comparisons to those who can be seen as using them with objectivity.

Even people with considerable technical training can have difficulty grappling with the implications of the data. How can people without any training be expected to understand?

While it is true that some people will never master the intricacies of technical data — and may never want to — concerned people can and do. People with little formal education have tackled reports of hundreds of pages, critiqued them, and even found errors when they felt their families' health was on the line. When people are motivated, they learn. It's a mistake to think that people can't understand; and it's an even bigger mistake to assume that those who disagree with you are ignorant.

5   TALKING WITH
THE MEDIA

♦ PLAN A PROACTIVE MEDIA STRATEGY

Planning a proactive media strategy means deciding what messages you would like to see in the media over the next several weeks (or months, or even years), determining what the company must do in order to receive the sort of coverage it wants, and then developing a program to achieve these objectives. To carry out an effective media strategy, moreover, a company needs more than just an understanding of media norms and constraints and the skills and resources to make company messages newsworthy. Media relations cannot be divorced from community relations or from still broader matters of company policy. How you are covered, in other words, depends on what you do. In companies with proactive media relations strategies, communicators do not work in isolation; they advise on media aspects of policy at the highest levels.

Not all companies believe it is wise to court public visibility with a proactive media strategy — at least not when the topic is environmental risk. Some argue that news coverage of risk inevitably tends to alarm the audience, that in terms of long-term company interests the least coverage is the best coverage, that the hallmark of a wise media strategy is to maintain a low profile. There is much truth to the judgment that media risk coverage tends to be alarmist (see page 13) and that even balanced risk coverage can alarm. But the costs of trying to avoid coverage are legion: loss of control over the story, which is likely to come out anyway; loss of credibility with journalists and the public, who are far more suspicious of secrecy than of publicity; loss of the opportunity to educate the media and their audience. On balance, more and more companies think it wiser to speak out.

Locked in hot local controversies over a waste pipeline into the Atlantic Ocean, a Superfund site, and a proposed new pharmaceutical plant, the last thing CIBA-GEIGY needed at its embattled Toms River, New Jersey, facility was a big media story on its Title III emissions. Nonetheless, Glenn Ruskin, who manages government affairs and communications at the Toms River site, put together an ambitious Title III media plan. Ruskin remembers thinking that "when I submit this stuff to the EPA, the reporters are certainly going to want to get their hands on it." It was a CIBA-GEIGY senior vice president who posed the key question to Ruskin: "Would I rather go out front with this information early on, or would I rather wait until a reporter physically went and tried to extract it from me. It became pretty clear to me that the best thing we could do would be to move out quickly and out front on it, and to put together this communications plan."

✔ Elements of a Proactive Media Strategy

1. **Goals** — What are you trying to accomplish with your media program? What do you want people to know, believe, or do as a result? "Media targets" like so many clippings or so many minutes on local television are not goals. What matters is what you want the clippings and newscasts to say.

2. **Audiences** — Which audiences do you especially want to reach through the media? Think about the effects of your media strategy on local government, fenceline neighbors, employees, customers, activists, and other key audiences.

3. **Media** — Which media are most appropriate for your goals and audiences, and for the newsworthiness of your story? Do you really stand a chance in the competition for TV time, or should you be aiming more at weeklies and radio talk shows?

4. **Messages** — What are the key messages you should stress in order to move your audiences toward your goals? One hallmark of proactive media relations is continually stressing the same points instead of getting seduced by the story of the moment.

5. **Concerns** — What are the long-term concerns of your key audiences? What suspicions or reservations must you alleviate to achieve your communication goals? The most effective media relations finds links between the interests of the source and the concerns of readers and viewers.

6. **Newsworthiness** — The trick to being newsworthy is to embed what you want to say in public events that are interesting, important, and easy to cover (see page 107), then to pare down the message so it is simple and focused (see page 84). If what you want to say is highly technical, this is a difficult trick indeed.

7. **Policy implications** — What you say should square with what you do. At a minimum, you must adjust your media strategy to match real performance. Better yet, work for improvements in performance that give the media strategy something solid to talk about.

8. **Logistics** — What resources will you need to carry out the program you have outlined? Who has to approve before you

can get started? What does a realistic timeline look like to achieve all that you have proposed? What are the appropriate first steps?

9. **Trouble-shooting** — What is likely to interfere with implementing your media strategy? What can you do now to prevent the preventable problems and prepare for the inevitable ones?

10. **Feedback** — You need feedback to keep your media strategy current (see page 136). Content analysis can tell you exactly what the media are saying about you; tracking polls can monitor changes in public opinion; even an informal chat with a few neighbors is better than nothing.

◆ PLAN YOUR MEDIA STRATEGY TO BE CONSISTENT WITH THE NORMS AND CONSTRAINTS OF JOURNALISM

Reporters and editors work neither for nor against the sources they cover. Their job, in a nutshell, is to find out what has happened — especially interesting, important, and recent things that have happened — and to package what they have found in an appropriate format for their audience. Journalists do not pursue ultimate truths, certainly not ultimate scientific truths; their focus is on events rather than principles, and their goals are accuracy and balance rather than "truth." They work under heavy constraints of time, space, and expertise.

The norms and constraints of journalism often lead the media to pay more attention to the alarming aspects of a risk story than to its reassuring aspects; warnings are both more interesting and, if true, more important than reassurances. In the interests of clarity and audience appeal, journalism similarly tends to polarize, simplify, and sometimes sensationalize risk information. Sources who understand the roots of these distortions are better able to prevent or mitigate them than sources who see them as incompetence or bias. A media strategy that is consistent with the norms and constraints of journalism stands the greatest chances of success.

Immediately after the catastrophic methyl isocyanate release at Union Carbide's plant in Bhopal, India, the company decided that all media communications should come from corporate headquarters in Connecticut. The decision to centralize information flow was designed to avoid inconsistencies and errors, but it nearly backfired when the media discovered that the only domestic methyl isocyanate facility was Carbide's plant in Institute, West Virginia. "So we immediately had national and international news coverage," explains regional public affairs director Thad Epps. "We had dozens of reporters and famous TV people and no one could talk to them." Epps, a chemical engineer, was working in government relations at the time. "I called Connecticut," he recalls, "and said, 'hey, someone has to talk to these people.'" Epps' comments wound up with the chairman of the board, who said, "'okay, but you have to do it....' So I became an instant media person." Though there are many good reasons for centralizing information flow, Epps' experience at Institute during the

Bhopal disaster suggests that too much centralization won't work, because the media need a fast, local response to a breaking story and are likely to interpret centralization as a kind of stonewalling.

Within 8 months of the Bhopal accident, the Institute plant had its own accidental release, not life-threatening, but certainly newsworthy — especially given that it was a Union Carbide facility. Epps realized the importance of telling the media what he knew as quickly as possible — even though he didn't yet have all the answers. Not wanting to let reporters into the plant, he held a news conference in the parking lot. The location turned out to be a mistake — "the heat was wretched," Epps recalls, and there were no boundaries, so cameras wound up surrounding the speakers. A follow-up news conference was held inside the plant, with better results. The lesson this time was that logistics matter — and part of crisis planning is deciding beforehand where you will meet the media in the event of an emergency.

Thomas McCollough of the Sun Refining and Marketing Company says that process is as important in communicating with journalists as it is with communities. Says McCollough, "What I've tried to do is establish some trust and credibility in a non-emotional event...so I have an open door policy." McCollough says that being open to talking to reporters when *they* need information — not simply when the company wants coverage — helps to establish him as a resource.

✔ What Makes a Good News Story?

With few exceptions, the media are not out to get you, just to get the story. But the characteristics of a good news story can sometimes lead journalists to act as if they were trying to "get you." The best protection is knowing what makes a story newsworthy and helping to frame the risk story at your facility so there is a good story available that *doesn't* get you. What then makes a good news story?

1. **Interesting** — The components of interest are many: drama, conflict, human interest, humor, surprise, etc. Emotional citizens are more interesting than impersonal corporate spokespeople. Examples are more interesting than data. And unless something is done to enliven the reassuring side of the story, high risk is intrinsically more interesting than low risk.

2. **Important** — Does the story affect the audience in significant ways? Does it involve powerful people or lots of people? The warning that industrial emissions may be hazardous is important, especially if leveled against top management. The reassurance that the emissions may be safe is less important, especially if handled by mid-level technical staff. Corporate sources must respond to stories reporters consider important and must build the importance of stories they want reporters to respond to.

3. **Timely** — What happened yesterday is enormously more newsworthy than what happened last month, and the race to get the story first is the essence of media competition. So is the race to find a new angle on yesterday's big story, or on this month's hot topic. If you can provide one, fine; if you can't, reporters will ask their questions elsewhere.

4. **Accessible** — Sources who are easy to cover get better coverage than sources who make life difficult. Returning reporters' telephone calls immediately, finding the information the reporter needs before deadline, scheduling news conferences for convenient times and places — these are the components of access. Journalists usually find activist groups very accessible, but complain that corporate management is hard to get to.

5. **Secret** — Reporters are always on the lookout for scandal, for illegal or unethical behavior in high places. Illegal or unethical behavior in not-so-high places is much less newsworthy, which is why a corporation that distorts data is much more vulnerable than an activist group that does likewise. The bottom line: keep as few secrets as possible.

6. **Simple** — Simple stories are easier for technically untrained reporters, editors, readers, and viewers to understand. The pursuit of simplicity helps explain why the politics, morality, and emotions of risk get better coverage than the science. Reduce your story to a few key concepts.

7. **Objective** — Journalists do not pursue "Truth" with a capital T. They don't think it is possible to know how hazardous industrial emissions "really are." Instead of "truth," journalists value objectivity, by which they mean giving both sides a chance to make their case—though not necessarily an equal chance (extreme views are more newsworthy than moderate ones, and alarming views are more newsworthy than reassuring ones). Sources do better once they stop trying to convince reporters not to listen to the other side.

8. **Accurate** — Journalists try hard, though not always successfully, not to mangle what their sources tell them, and they greatly resent sources who feed them misinformation. A debatable (or even outrageous) opinion, on the other hand, does not constitute misinformation. For journalists, the story is accurate if it faithfully reproduces the views of the contenders. A source's bottom-line job, in other words, is to get the facts right; a journalist's bottom-line job is to get the quotes right.

9. **Concrete** — News is about events — especially events that are recent and public, interesting and important, local and unusual. It is not about scientific principles or universal truths or abstractions of any sort. The essence of media relations is to embed the issues and information you want covered in events that make them newsworthy. Activist groups often do this with flair; corporations often neglect to do it at all.

◆ COORDINATE YOUR MEDIA RELATIONS STRATEGY WITH YOUR COMMUNITY RELATIONS STRATEGY

Some company risk communicators understand the need to talk directly with affected groups, but hate to deal with reporters. Others may enjoy working with journalists, but shy away from direct contact with the local community. Unfortunately, neither of these halfway approaches is effective. Good community relations and good media relations rely on each other.

The most avid readers and viewers of media risk stories are the people who are most directly affected — company employees, local government officials, fenceline neighbors, environmental activists, people who are (or at least consider themselves) *stakeholders* in the company's risk management activities. They check the news in search of additional information and confirmation of what they already know. They are displeased if what the company is telling the media is different from what it is telling them. And they are even more displeased if the company is telling the media things they feel it should have told them first.

When companies do provide information to stakeholders first, unfortunately, one of the stakeholders may well release it before the company is ready. Especially if the news is important or negative, then, coordinating community relations and media relations can be a matter of careful timing.

At Dow Chemical, for example, an epidemiological study on plant benzene emissions was due to be released. First to be alerted were the technical people who had done the research. Next came key government officials at the Environmental Protection Agency, the Occupational Safety and Health Administration, the state Department of Natural Resources, etc. Next in line were key top managers at Dow and Dow employees who were directly affected (those whose occupational exposures had figured in the study, for example). Outside experts came next on the contact list — so reporters would have someone to call for third-party validation of the results. Finally, unaffected employees and the media were contacted.

Of course, as Dow community relations supervisor Susan N. Dupree points out, the sequence must be tailored to fit the specific situation. And speed is more important when the news is uncertain or bad than when it is favorable. But in all cases, says Dupree, the stakeholders who might otherwise feel blindsided are informed first, then unaffected employees and the media.

Explains Dow's Colin Park, manager for issues management/ biostatistics, "We get half a dozen people together and we noodle through the scenario of how the data is going to come out, and then make sure that that's the way we want it to happen. 'Gee, should we tell our critics, should we tell the environmentalists?' 'Yeah, they ought to hear at the same time DNR hears and OSHA hears.'" Being sure nobody is blindsided by the sudden appearance of a media story, says Park, allows Dow to shape the discussion more. At times it can even "turn potential adversaries into supporters."

✔ Integrating Media Relations and Community Relations

WARNING: The following list of things to think about before scheduling a news conference should not be used as an excuse for failing to deal with reporters. The media *will* get the story, most of the time, whether they get it from you or from your opposition. Staying on top of the story is essential, and that means moving fast; but it doesn't mean ignoring the interpersonal contacts that should precede most media contacts.

1.   **Internal coordination** — Have you forewarned those in your company who are likely to be called by the media in pursuit of the story, or who should see or approve the story in advance for any other reason?

2.   **Next steps** — Have you thought through what may happen after the story breaks? If you expect continuing media attention, are you ready to cope with reporters scrounging for fresh leads and comprehensible background? What sorts of concerns are readers and viewers likely to have? Customers? Employees? Neighbors? Figure out who will have to field most of the questions and make sure they are ready with the answers.

3.   **Employees** — Have you alerted employees who are affected? Employees need information to answer their own questions and their friends' and neighbors' questions; and don't forget retirees and the families of employees.

4.   **Government** — Have you alerted government officials (elected and appointed) who may be contacted by the media, or by their constituents after the story breaks? Officials should be briefed so that they respond appropriately and don't feel blindsided; but the briefing must be tactful, so they don't feel pressure to speak for your company.

5.   **Involved parties** — Have you alerted those who have previously established their interest in the issue — environmental activist groups, for example? Often it is helpful to brief even your opponents; when a company is in fact behaving responsibly, the company is harder to attack with information than without.

6. **Neighbors** — Have you alerted the nearest neighbors of your facility, the people who are most likely to be affected by any risk you generate? If you have new data about effluent releases, for example, your fenceline neighbors should hear about it from you, not from a reporter or a news story.

7. **The business community** — Have you alerted others with whom you do business — major customers, for example, or nearby companies with similar operations involving similar risks? A joint announcement may or may not be appropriate, but customers and competitors probably do need time to prepare for the questions your announcement may generate.

8. **Experts** — Have you alerted the local experts of greatest relevance — university environmental engineers, hospital toxicologists, perhaps even the local high school science teacher? Sooner or later, most experts will be asked for their opinions, and their answers may depend, in part, on their having complete information.

9. **The toughest calls** — Who are you particularly avoiding because you are worried the contact may be unpleasant? In the long run, the company is probably not better off continuing to ignore them. Consider gritting your teeth and making the calls.

10. **Other stakeholders** — Who else should you contact in advance? Who else has a stake in the outcome? Who else is likely to be contacted by the media or contacted by others after the story appears? Who else has a history of involvement in the issue? Who else has relevant expertise and may be asked for an opinion? Who else will feel slighted, ignored, or blindsided if the story appears without forewarning?

♦ KEEP YOUR MEDIA MESSAGES FOCUSED, SIMPLE, AND CLEAR

Most reporters (though not all) have very little technical background; the same is true of most editors and most readers and viewers. Most reporters, editors, readers, and viewers, furthermore, are convinced that the essence of the risk story is not technical.

They are largely right. While a corporate expert may be eager to explain why the risk of some chemical emission is only 10^{-6}, not 10^{-5} as a state regulatory agency may be claiming, the reporter and the public are much more interested in what the company proposes to do to get chemical emissions out of the air and water and why it didn't take action years ago. Some technical information is usually essential to the story, of course, but seldom as much as the technical specialists suppose (see page 107). A wise source, therefore, keeps the news release or interview as nontechnical as possible. A wise source, furthermore, works hard to make those few technical points that cannot be dispensed with simple, clear, and interesting.

Perhaps most important, a wise source *chooses* what points to make, what evidence to provide, what public concerns to address — what to say. Many inexperienced interviewees are curiously passive in their contacts with journalists, feeling that their job is simply to show up and answer whatever questions the reporter puts to them. Of course it is important to answer the reporter's questions; but it is equally important to anticipate the reporter's questions and to decide in advance how to use your answers to make the points you want to make.

Television, of course, requires a more pared-down approach than print. Even a newspaper reporter must simplify, but print permits the reporter space to explain and time to figure out what needs explaining. By contrast, broadcast journalism relies on 10- to-20-second "sound bites," self-contained statements of no more than a few pithy sentences. Seasoned broadcast news sources talk in sound bites and prepare their sound bites in advance.

Experienced sources on all sides of risk controversies agree that media messages should be kept focused, simple, clear, and largely nontechnical. Geraldine Cox of the Chemical Manufacturers Association urges scientists to "speak in headlines," while Ellen Silbergeld of the Environmental Defense Fund says it is essential to explain things so that "your mother can understand it." Advises Ralph H. Hazel of the U.S. Environmental Protection Agency: "Don't go into an interview with a huge suitcase of minute details. Go into an interview with an outline in your head of the important points you want to make."*

---

* From Sandman, P. M., D. B. Sachsman, and M. R. Greenberg, *The Environmental News Source: Providing Environmental Risk Information to the Media* (Chelsea, MI: Lewis, in press).

✔ Identifying the Key Points to Make in an Interview

Most experienced sources simply refuse to begin an interview or take a telephone call from a reporter until they have spent five minutes identifying the key points they want to focus on. Picking *three or four key points* — not twenty — doesn't guarantee that the interview will go your way, but it improves substantially on the odds. The following questions will help you choose and develop your main interview themes.

1. What do you especially want the public to know, understand, or believe?

2. What is your best evidence that the items in #1 are so?

3. Why would some people disagree about the items in #1? What are their key arguments? What are your answers?

4. What are the public's key concerns? How can you respond to these — not rebut them, but truly *respond*?

5. To what extent are #1 and #4 connected? If they are not closely connected — if what you want people to know isn't responsive to their concerns — consider revising #1.

6. What else is the reporter likely to ask? How will you respond? How can you connect your answers to #1 or #4?

7. What else do you need to explain in order for people to understand your answers? What background must you provide? What misconceptions must you clear up? What misconceptions should you be especially careful to avoid creating?

8. What else are you tempted to talk about? Is there really a reason to include it, or is it a technical detail that the reporter and audience can manage without?

9. How can you make your answers newsworthy? How can you make them interesting, punchy, and graphic? How can you make them personal? How can you make them simple and clear?

10. Of the talking points that emerge from these nine questions, which three or four do you most want to see in the news story? These are your key points. Stress them every chance you get.

✔ How to Help Reporters Understand a Technical Story

1. **Don't assume knowledge** — Unless you know otherwise, assume that the reporter is smart but technically ignorant. This is so even if the reporter is nodding vigorously in apparent comprehension; some journalists hide their ignorance behind a polished mask of professional bravado.

2. **Guide the interview** — Know what points you think are key, and keep stressing them. If the reporter seems to be going in what you consider a mistaken direction, tactfully point out that you think X is much more central to the situation than Y.

3. **Avoid jargon** — If there are technical terms that you absolutely must use, define them carefully — especially if they are terms that the media often use incorrectly; but for the most part don't use them at all.

4. **Simplify content** — If you don't, the reporter will have to simplify for you—and you are more qualified to decide which details are dispensable. But be careful not to oversimplify; in particular, never leave out a point whose omission will make your interview look misleading or self-serving when someone else raises it.

5. **Anticipate problem areas** — Think through in advance the likeliest ways reporters may misunderstand — concepts other reporters have got wrong in the past, areas where your thinking is counterintuitive or different from the industry stereotype, etc. Make a point of explaining that people often think you mean X, but what you are really saying is Y.

6. **Provide written back-up information** — Use glossaries of technical terms, summaries of key data, chemical profiles, etc. If the interview is by telephone, offer to fax the relevant documents to the newsroom.

7. **Be alert for signs of confusion** — When confused, the reporter may stop taking notes, or take notes furiously, or follow up your answer with what sounds like a nonsequitur,

---

* From Sandman, P. M., D. B. Sachsman, and M. R. Greenberg, *The Environmental News Source: Providing Environmental Risk Information to the Media* (Chelsea, MI: Lewis, in press).

or begin to look glassy-eyed or nervous. When you suspect that the reporter is lost, back up and make your point again.

8.   **Check for understanding** — Geraldine Cox of the Chemical Manufacturers Association asks reporters to try paraphrasing her point. Other sources telephone reporters after the interview to ask if they have any questions. Be tactful: you are not trying to catch the reporter in an error, but rather trouble-shooting your own ability to communicate complex information.

9.   **Suggest other sources** — It is entirely appropriate to suggest that the reporter interview other sources who are likely to confirm your own position, especially if they can do so with more expertise or more credibility. But it is often wise to refer the reporter to responsible sources with opposing views as well.

♦ WORK HARD TO PROTECT AND ENHANCE YOUR CREDIBILITY WITH THE MEDIA

Credibility is easy to lose and hard to regain. It is also precious; protecting and enhancing your credibility is essential to effective media relations.

Of course, many sources have their own reservations about the performance of journalism — bias, inaccuracy, sensationalism, and incompleteness are the most frequent complaints. But the situation is not symmetrical. Reporters take solace from the fact that sources on all sides disparage their work, figuring that universal disapproval is a sign of successful neutrality. But they pick and choose among sources largely in terms of their credibility.

Journalists do not demand that a source be "objective" in order to be credible. Most sources represent a point of view, and to some extent what they say is inevitably and appropriately filtered through that point of view. But journalists do expect a source to be honest, accurate, helpful, accessible, and open — a difficult task when the information being sought is itself frought with uncertainty.

The job is especially difficult for corporate spokespeople representing smokestack industries or other generators of environmental risk and environmental controversy. Even more than with other sources, journalists (and the public) tend to suspect that a company representative may distort or misrepresent the truth in pursuit of company self-interest. Opinions differ on whether this judgment results from the less-than-honest track record of corporations, the liberal bias of journalists, or other factors, but it is widely shared. You start at a credibility disadvantage.

"When I talk with a reporter," says Carl Patrick of the New York Power Authority, "I try to distinguish the industry position on something as opposed to the scientific fact behind it. When I say this level of radiation is acceptable, that's an industry position, and you may find somebody else who will dispute that. But we find ourselves being quoted that the industry contends that purple can be formed by mixing red and blue — which is not a contention but a scientific fact."*

---

* From Sandman, P. M., D. B. Sachsman, and M. R. Greenberg, *The Environmental News Source: Providing Environmental Risk Information to the Media* (Chelsea, MI: Lewis, in press).

While industry spokespeople watch their factual statements devalued as mere opinions, academic experts enjoy the opposite experience: their opinions, even outside their fields, are often quoted as facts.

✔ Tips for Protecting and Maintaining Source Credibility

1. **Understand and respect the journalist's job** — You gain credibility if you show you understand the norms and constraints of journalism (see page 105). You lose credibility if you seem contemptuous or patronizing.

2. **Be honest** — A source who loses credibility because of an error will eventually regain it, but a source who willfully misleads the media will never be trusted again.

3. **Be accurate** — Check and double-check your facts. If you're not sure, don't guess; tell the reporter that you're not sure but you'll find out, and then do so — by deadline.

4. **Avoid extreme statements** — A source with extreme opinions will earn plenty of coverage, but not much credibility — especially if the source has to back off a few of them when confronted with conflicting evidence.

5. **Avoid extreme caution** — A source who never ventures much beyond "further research is needed" may be credible enough, but he or she is of little use to the media. Have something to say.

6. **Be helpful** — Sources are judged more credible when they are cooperative, when they manage the interview logistics efficiently, return telephone calls quickly, etc.

7. **Be accessible** — Be there when reporters need you on deadline. And make sure the other key players at the company are available as well; reporters especially want to talk to the top manager (who makes the decisions) and the technical expert (who knows the answers).

8. **Avoid keeping secrets** — You may have good reasons for not being completely open — anything from quality control to liability suits — but the decision to keep mum will cost you considerable credibility. If you have to withhold information, explain why, and say when the information will become available.

9. **Avoid peddling fluff** — While journalists like sources to be open, they don't want to waste a lot of time skimming unpublishable releases or attending unnewsworthy news confer-

ences. Sources who try to peddle a lot of bad stories have trouble selling a good one when it comes along.

10. **Avoid peddling the "industry line"** — Reporters don't object when industry spokespeople represent the industry view; in fact, they call you for the industry view. But predictable, defensive, reflex boilerplate is a lot less credible (and less newsworthy) than an individual, thoughtful response to a particular situation.

11. **Provide background** — You can measure your credibility by how often a reporter asks you to fill in the background or to suggest people to interview — even when you're not in the story yourself. You can build credibility by volunteering this sort of information.

12. **Be personal** — Make your interview answers personal; especially when talking about illnesses or injuries, remember that these are tragedies, not just incidents or data. Build a personal relationship with the reporter as well, as you would with any other colleague with whom you anticipated continuing contact.

13. **Aim for consistency** — Two kinds of inconsistency cost sources credibility: changing spokespeople too often and having too many spokespeople providing incompatible information. But don't centralize information flow too greatly — reporters resent that too.

14. **Be energetic and enthusiastic** — It may be unfair, but lively sources are generally seen as more credible than withdrawn ones — and they are certainly more effective.

15. **Be forgiving** — Whether or not they forgive your mistakes, journalists expect you to forgive theirs. Friendly feedback — praise as well as criticism — goes further than angry demands for a correction. When it is necessary to correct the record, find ways to do so without raising the roof — and limit your complaints to the publisher or station manager to one per decade.

Yes, but...

Going to reporters with a story that has bad news is unwise. We are better off hoping we can duck the issue.

The down side of this approach is that if you bet wrong, some reporter will come to you anyway. You will have less control of the story, be more apt to be on the defensive, and run a risk of losing credibility. This approach also reinforces the general perception that industry is not to be trusted.

Media coverage is so biased it makes sense to avoid it whenever possible.

It is true that reporters have very different ways of seeing a story than many in industry do; but trying to avoid reporters will often lead to worse coverage. In many cases, you are better off trying to give your perspective than abdicating to other sources. The story that says your company refused comment can be the worst of all.

6   SETTING THE STAGE

◆ DEVELOP INTERNAL PROCEDURES AND ALLOCATE RESOURCES TO ENCOURAGE EFFECTIVE RISK COMMUNICATION

Companies seeking to improve their risk communication should examine not only their interactions with audiences outside their gates, but also their internal interactions to determine how they encourage — or discourage — risk communication.

Industry practitioners assert again and again that risk communication is "internally driven" — the result of the efforts of qualified staff encouraged by the commitment of senior management to risk communication. Unfortunately, in many cases this commitment has come after the company was embroiled in a high-profile environmental controversy and was forced to deal with outside interests. Although crises can provide powerful incentives, avoiding risk communication disasters is far less resource-intensive than mopping up after them.

Proactive risk communication — the effort to avert communication crises — is difficult without management commitment at the corporate or at least the plant level. Staff *can* initiate risk communication efforts and can bring risk communication to the attention of management. But staff efforts supported by upper management will be much easier to implement and are much more likely to succeed. Because risk communication is usually inextricably tied to risk management issues, cooperation — if not active participation — is often needed from a variety of people within the organization.

Keith Fulton of Exxon Chemical stresses the tie between risk communication and senior managers: "Senior management has to say to the manufacturing management...'I want you to personally spend 10 to 25% of your time developing, nurturing, and participating in these activities, and I want you to tell me those people in your organization that you think will need to spend more time.'" He emphasizes that without management commitment, communication efforts are likely to become mere "tokenism."

Fulton makes the point that when management places a priority on risk communication, resources tend to fall into place: "Perhaps what we have done is recognize that we don't have any choice.... If you take that mentality about this, then you go and do it and drop something else. All of us have more to do than we can possibly do. So the attitude that 'I don't have the resources to do it' is just false."

Scarlett Lee Foster of Monsanto argues that communication successes can serve as models and encourage other efforts — even in the absence of a clear mandate from senior management. She notes that a plant manager who coped with dioxin contamination near the plant by releasing information quickly and dealing with the situation aggressively was "driven by necessity." According to Foster, "He knew he had a lawsuit coming up, this dioxin thing came out of the blue and then Bhopal came out of the blue.... His ability to get through this unscathed prompted other plant managers to go out and be more proactive, at least build a base of trust.... But, it really varies by plant manager."

✔ How Management Can Encourage Effective Risk Communication

When top management encourages risk communication, staff and mid-level managers will be less reluctant to take on this often difficult task. Although good communication skills are essential for effective risk communication, skills are only part of the equation. The following suggestions from practitioners are steps in the right direction.

1. **Practice effective risk management** — When a company manages risks well, meaningful risk communication is possible. Dialogue with outside interests may further improve risk management procedures, but no amount of risk communication can cover up mediocre environmental practices.

2. **Make a commitment** — The attitudes of senior management toward environmental concerns and those who raise them is critical to effective risk communication. Without meaningful commitment from senior management, risk communication is unlikely to occur consistently. Senior managers should model good risk communication and indicate its importance by acting as spokespeople at key meetings.

3. **Open internal communication** — If staff members are encouraged to contribute dissenting ideas at internal meetings, they are more likely to feel comfortable dealing with dissent outside the company. Procedures are needed that reward a manager who airs problems while they are small — raising problems before they become a long-term liability.

4. **Allocate resources** — Risk communication should be treated as a necessity rather than an expendable luxury. The more controversy there is likely to be, the more resources are likely to be needed. Hire communication staff who have experience not only with developing written materials but also with encouraging dialogue.

5. **Develop systems to amplify public concerns** — Staff who deal with the public should have means to alert more senior managers to outside voices before those voices feel a need to shout. Managers should avoid asking staff to serve as buffers between them and public concerns; instead, they should work to institutionalize the amplifier role.

6. **Train managers and staff** — Managers (including senior managers), technical staff who are likely to interact with the public, and public relations staff should receive training that not only builds communication skills (emphasizing two-way communication), but also explains how the public perceives risk and how the company can affect that perception.

7. **Develop models** — Sharing information about effective risk communication can encourage others to try similar efforts. Management should look for models to disseminate and ways to disseminate them, such as "good newsletters" and in-house seminars.

8. **Provide incentives to communicate** — Communication should be built into job descriptions and personnel evaluations. Efforts to avert communication crises should receive as much, if not more, recognition as efforts to deal with crises.

9. **Develop risk communication plans** — Communication efforts that are carefully planned tend to be more useful and less resource-intensive in the long run. Company planning documents dealing with environmental issues should also consider communication issues.

10. **Evaluate your efforts** — Communication efforts, like technical ones, improve with feedback (see page 136).

✔ How to Persuade Managers to Support Risk Communication

Although a company's risk communication activities are determined in large part by the tone set by senior managers, on a day-to-day basis line staff and mid-level managers vitally affect how the company communicates. While it is easier to work with a mandate from the top, risk communication efforts have succeeded through the efforts of mid-level managers and public relations staff. Even when there is a mandate from the top, rarely is implementation smooth, and risk communication advocates are needed at *all* levels. The following are some suggestions for moving risk communication upwards in your organization. Some of the suggestions are similar to those listed in "How Management Can Encourage Effective Risk Communication" because in some instances encouraging risk communication will look the same whether it is from the top down or the bottom up.

1. **Provide information on successes** — Sometimes risk communication successes can help others overcome their reluctance. Newspaper articles, summaries, journal articles, etc. which describe risk communication efforts that succeeded may bolster the case for yours.

2. **Evaluate efforts** — Feedback — particularly quantitative data — concerning risk communication efforts can be particularly useful to obtain support for subsequent efforts (see page 136).

3. **Collect other data** — Information from newspaper coverage, polls reported in the popular press, journal articles, etc. concerning public attitudes and concerns may be useful to support your ideas.

4. **Forward information** — Pass on articles and information about other companies' controversies with suggestions for communication approaches that they should have used and that your company should use now to avoid problems.

5. **Collect evidence** — Draw on your own documentation that risk communication works — letters of approval from citizens, for example — and pass them up the hierarchy.

6. **Provide concrete steps** — Some managers may have diffi-

culty endorsing risk communication as an abstraction, but may find well-thought-out, easy-to-implement, concrete steps more appealing.

7. **Improve your skills** — If your company won't pay for training, read current publications in the field for ideas, ways to avoid common pitfalls, etc. (see "Suggestions for Further Reading").

8. **Network** — Outside support and feedback help, especially if they are hard to find within the company. Many professional meetings dealing with environmental subjects now have risk communication sessions involving more experienced practitioners.

♦ CHOOSE CAREFULLY WHO WILL REPRESENT THE COMPANY

Dealing with the public can be one of the toughest jobs companies face. It can be easier to manage industrial risks than to explain them. However, if a company does a superb job of managing risks and a poor job of explaining them, the company is unlikely to seem credible. (As pointed out earlier [see page 19], the converse does not hold true: good risk communication cannot compensate for mediocre risk management.) Companies need to put the same care into selecting people to deal with the public as they put into selecting those who manage risks.

Ideally, a company should choose as a spokesperson someone who has a good feel for both the technical data and dealing with public concerns. When in doubt about who to assign to deal with the public, a company is probably better off with a spokesperson with a strong technical background who is "good with people" than a PR professional with no technical background.

In many situations, such as meetings, a team approach may work best, with various technical people fielding questions related to their disciplines. Dow Chemical Company, for example, uses such an approach by having communication people do behind-the-scenes planning, while technical staff who know their subject in depth respond to questions and concerns. Such a team approach requires advance coordination, a good chairperson, and communication training for the technical staff.

Although some companies do use lawyers as spokespeople, lawyers — by virtue of their position in the company — may give the impression of a company that is more concerned about the letter of the law than people's welfare. A company that leads with a lawyer on anything but clearly legal issues is running the risk of being seen as uncaring, particularly if the lawyer acts reluctant to admit too much because of concerns about liability.

Exxon Chemical selects spokespeople based on the jobs they're in and fills these jobs with people who can handle the spokesperson role. As Keith Fulton of Exxon Chemical puts it, "We do consciously say, 'This person will be in the public,' and they clearly have to have certain skills or better develop them real fast." To meet this need, provides people with both media and public speaking training. The company also keeps these "spokespeople-in-training" together, "... so we're self-training each other."

Joan Ebzery, director of public affairs for Clean Sites, Inc., stresses the importance of having technical people address public meetings. "While I can have the facts, I don't have...the technical depth that the technical people do." Instead, she sees her role as "to make sure that people get a lot of information and get it in a timely fashion...[to] put them in touch with the people who...they can talk to." Ebzery notes that she gives a lot of routine information to the people who "don't want to talk to the person who knows PCBs." She also emphasizes the importance of having spokespeople who are based in the local area and are available when people need them for follow-up.

✔ Attributes of a Successful Spokesperson

If you look at the "great communicators," their greatness often comes from their ability to seem human — even when those around them are retreating behind their roles. Technical mastery or glibness will take spokespeople only so far. Although laypeople may not be able to discern technical nuance, they are usually quite expert at detecting arrogance or defensiveness.

While great communicators may be born, most others can be improved with training, experience, and feedback. The following are some attributes of a good spokesperson, and, within limits, all of them can be cultivated. Ideally, a company should pick people with many of these attributes and then help them build their strengths and overcome their limitations.

1.   **Confident** — Spokespeople should have confidence in their technical knowledge, their ability to respond to a range of concerns, and their speaking skills.

2.   **Prepared** — Spokespeople should be clear about their "bottom line" messages and be ready for likely questions.

3.   **Technically competent** — Don't send a PR person to respond to technical questions.

4.   **Down-to-earth** — As Stanley Dombrowski of Dow puts it, spokespeople should be good "laymanizers." Concrete information that addresses the audience's concerns often works best.

5.   **Accountable, accessible, and local** — A spokesperson should be rooted in the community — not a commuter.

6.   **Authoritative** — People want to know that the company cares enough to send someone who has sufficient authority to "speak for the company." The best combination in a high-controversy situation may be a local manager *and* a boss from out of town.

7.   **Empathic** — When people have concerns, they want to deal with someone who can share them. The ability to listen is one of the most important and yet most overlooked communication skills. In conflict situations, hearing others' points of view is often more critical than the comebacks.

8. **Spontaneous** — Because communications plans often need to be changed, it helps to have spokespeople who can go with the flow and think on their feet.

9. **Forthright** — Candor goes a long way toward building trust. Nibbling around the edges of the truth will gouge credibility (see pages 37 to 42).

10. **Centered** — When others become emotional, good spokespeople do best by remaining calmly caring.

11. **Articulate** — Obviously.

✔ How to Use Public Relations Professionals

Public relations professionals are too often hampered by the conflicting attitudes at the companies they work for, whether as consultants or in-house staff. On the one hand, they are sometimes asked to help companies make Houdini-like escapes from tight situations and magically provide happier endings for public relations disasters. On the other hand, they may be treated, unlike those with technical background, as if they had no particular expertise; those with technical backgrounds may feel no compunction about overruling their advice. This makes it much more difficult for public relations professionals to help companies improve communications performance. Following is a list of ways companies might profitably use public relations professionals.

1. **Guidance** — Instead of calling in the public relations professionals *after* actions and policies have been determined, it is more useful to seek their advice during the decision-making process.

2. **Communication strategy** — Effective communication planning will maximize impact and resources. Anticipating communication needs and planning to avoid problems work much better than waiting to respond to them.

3. **Liaison** — Public relations professionals are useful to serve as links to communities, environmentalists, and other outside groups. They can also bring community concerns back to the company.

4. **Training** — Public relations professionals with training background should be called upon to provide spokespeople with training in handling the press, making presentations, dealing with communities, etc.

5. **Feedback** — Not only should public relations professionals provide formal training, but they can also provide valuable feedback on an ongoing basis. If they don't understand the handout or "buy" the presentation, the public isn't likely to either. Debriefing should be standard operating procedure after public meetings.

6. **Writing** — Good public relations professionals should have

outstanding instincts concerning what will be understandable to laypeople. They are excellent translators of technical material.

7. **Facilitation** — Public relations professionals can sometimes be useful to chair public meetings, structure agendas, keep meetings on track, and channel questions to the appropriate technical person.

8. **Evaluation** — Public relations professionals can help gather data to guide the company's risk communication decision making (see page 136).

✔ How to Torture Public Relations Professionals and Undermine Your Company's Credibility

As the phones ring off the hooks, the message slips cover desk tops, and the cloud of public opinion descends, odds are good that public relations professionals will be asked to make it all disappear...quickly. Odds are also good that they might have seen it coming but were tortured into silence. Next time the public relations troops need to be called in to mop up a disaster, see if the company has gotten there by using public relations professionals in the following ways and if the company will potentially escalate the conflict by abusing them further.

1. **Ping-pong ball** — Require them to obtain 99 different approvals for a news release while reporters wait.
2. **Buffer** — Ask them to make sure senior managers are never bothered with trivial concerns — like mothers with kids.
3. **Apologist** — Tell them to make excuses instead of promises they can keep.
4. **All-purpose spokesperson** — In ten minutes anybody can learn toxicology.
5. **Paper pusher** — When in doubt, make them write.
6. **Time buyer** — When disaster looms, have them say, "It ain't so."
7. **Miracle worker** — When disaster strikes, ask them to stop speeding bullets while explaining the trajectories.
8. **Fall guy** — When disaster strikes, say they did it.
9. **Flunky** — Ignore their suggestions. Anybody can do PR.

♦ EVALUATE YOUR EFFORTS TO COMMUNICATE WITH COMMUNITIES SO YOU CAN IMPROVE THEM*

Obtaining feedback on risk communication efforts can make it easier for companies to repeat successes and avoid pitfalls. Feedback can be particularly useful if it is used not only to assess the end result of a communication effort, but also to provide information for mid-course corrections. This can help conserve resources, preserve credibility, and improve dialogue with communities. In essence, feedback can help practitioners know when they are about to light communications fires, as well as guide their firefighting efforts.

Just as important, feedback can be useful to document the progress of communication efforts so that program assessments go beyond the hunches and gut responses of company representatives. Documentation can be essential to make a case for more resources or to secure support for a communication effort. Finally, when risk communication is seen as a two-way process, soliciting feedback from those with whom you are communicating becomes a crucial element. Asking people what they think of your communication efforts can be part of keeping an open channel. Obtaining feedback can become integral to communication rather than an "add-on." As companies make evaluation routine, it will become easier — and should improve their learning curve more quickly.

Monsanto began conducting telephone surveys in 1986 with a 1200-person sample of the community surrounding the company's plant in Nitro, West Virginia, where there were concerns about dioxin contamination. The survey, which was pretested, asked both closed- and open-ended questions about people's responses to the company, including, "What's the worst thing about the plant." Despite some fairly high-profile activity related to dioxin, Monsanto did well, with an 84% approval rating from the citizens in Nitro, according to Scarlett Lee Foster. This community will be resurveyed every 2 years to give Monsanto a better feel for citizen concerns. The survey was

* This section is derived in part from Kline, M., C. Chess, and P. M. Sandman, *Evaluating Risk Communication Programs: A Catalogue of Quick and Easy Feedback Methods* New Brunswick, NJ: Environmental Communication Research Program, Rutgers University, 1989).

designed so that it could be used in other communities, with Monsanto planning to resurvey communities on a regular basis and to expand each year to a new facility.

✔ Quick and Easy Feedback Methods

When most people think of evaluation, they think of approaches that give an overall assessment of the efforts of a program. Such rigorous approaches are usually sufficiently resource-intensive that they are most useful for large-scale resource-intensive communication efforts.

However, "quick and easy" feedback methods can be useful for small-scale efforts for which rigorous evaluation is not possible. For detailed summaries of each of the approaches listed below, including their strengths and limitations, please refer to *Evaluating Risk Communication Programs: A Catalogue of Quick and Easy Feedback Methods* (see Suggestions for Further Reading), which includes additional methods of obtaining feedback.

1. **Audience information needs assessment** — Questions from relevant audiences are gathered in advance of public meetings so a response can be organized and presented.

2. **Public opinion polling** — Polling can give companies a sense of public attitudes and perceptions so the company can better target its communications. Although polling can be an extensive undertaking, it can also be scaled to the size of the communication effort.

3. **SMOG readability grading formula** — Reviewing a sample of text from a written piece and performing some simple mathematical calculations will yield a SMOG grade, which represents the reading grade level a person must have reached to understand the text.

4. **Self-administered pretest questionnaires** — Questionnaires about written materials are developed to get feedback from people similar to the intended audience.

5. **Focus groups** — A discussion session run by a trained moderator is conducted to get feedback and generate ideas about pretest items.

6. **Meeting reaction form** — A brief form can be developed for distribution at public meetings to get a sense of whether information was transmitted, whether those in the audience felt their concerns were understood, etc. (see the sample Evaluation Checklist form on page 140).

7. **Observation and debriefing** — This is an easy, but often-overlooked way of getting feedback on a speech from neutral observers. Although useful, it's no substitute for finding out directly the audience's reaction.

8. **Assessment of communicator style** — There are a number of different tools to help communicators examine what they bring to the communication process. Most are self-assessment surveys that are most effectively used in combination with a consultant who can help communicators turn the results into behavioral changes.

Evaluation Checklist

Date: _____ Group: _____ Meeting Topic: _____

Our company is very interested in knowing what you thought of this meeting so we can do better next time. Please complete this survey before leaving to help in this effort.

1. How did you hear of this meeting?

2. Respond to the following statements using a scale of 1—5, where:

1 = agree strongly	2 = agree moderately	3 = neither agree nor disagree	4 = disagree moderately	5 = disagree strongly

   a. I had all my major questions answered in this meeting.   _____
   b. I learned a lot about the issues covered in this meeting.   _____
   c. Company representatives seemed to listen carefully to the opinions and questions of those outside of the company.   _____
   d. Company representatives were difficult to understand.   _____
   e. Company representatives seemed to speak honestly.   _____
   f. Company representatives did not deal with the issues that concerned *me*.   _____
   g. Company representatives dealt with the hard questions during this meeting.   _____
   h. Company representatives were unclear about their actions and plans.   _____
   i. Company representatives understood my feelings about this issue.   _____
   j. I believe the company will use input from this meeting in its decisions.   _____
   k. Company representatives seemed authorized to speak for the company.   _____
   l. I gained a better appreciation of the dilemmas involved in this topic.   _____
   m. Arrangements for this meeting (selection of time and place, directions, agenda, materials) were well-handled.   _____
   n. I feel a need for more meetings.   _____

3. The thing I liked most about this meeting was:

4. The thing I liked least about this meeting was:

5. Please use the back of this sheet for other comments, questions, or concerns.

   If you have additional questions, please contact _____ at _____ .

Yes, but...

We don't have the time or resources to think about communication unless there is a crisis.

As with environmental management, scrimping on planning may lead to communication crises that could have been averted; and dealing with communication crises is usually high-risk and labor-intensive. In many cases, investing time in planning will pay off by reducing resources needed later on. So can training staff and providing consistent feedback.

The bottom line: You can't afford not to communicate.

There is barely enough time to think about our communication, let alone to evaluate it.

If you don't solicit feedback, you may lose time by going off course. Evaluation can ultimately save time by speeding the learning curve.

Why should I take on a communication problem? If I wait, it may not explode until it is someone else's responsibility. But if it explodes while I'm handling it, it will definitely be my problem.

In fact, it is often in a company's interest to do proactive communication, but not so clearly in the interest of individual managers. This is a painful truth that calls for clear mandates to tackle tough problems before they become intractable. Companies will also need to develop incentives so that it is in managers' interests to do proactive communication.

7 TEN WAYS TO AVOID COMMUNICATING ABOUT RISK

Risk communication is risky. If you are ever tempted to take the easy way out, consider the time-honored techniques in the tongue-in-cheek list below. Better yet, mix and match this list with "Ten Ways to Lose Trust and Credibility," since avoiding communication and losing trust are inextricably linked.

1. **Talk instead of acting** — Rather than managing the risk, talk around it in hopes that people won't notice.
2. **Trivialize the risk** — Compare the risk to rolls of toilet paper or peanut butter to let people know you really care.
3. **Hide behind lawyers** — They can help you find reasons not to talk.
4. **If you make a mistake, deny it** — Never let people know you learned from your mistakes.
5. **Don't speak plain English** — Use "techno-babble" and bury key points in mounds of detail; or simplify so completely that you leave out important information.
6. **Lecture** — Make people wait to have their questions answered until you have given them the "important" information.
7. **Don't let down your guard** — Never let people see you are human. When people are upset, tell them they are "irrational."

8. **Wait to talk until someone else does** — Then act defensive. Blame the media and environmental groups for anti-industry bias.
9. **Scrimp on resources for communication** — Talk is cheap.
10. **Wing it** — Don't plan ahead or coordinate with others in your company. After all, communicating is easy.

SUGGESTIONS FOR FURTHER READING

Covello, V. T., P. M. Sandman, and P. Slovic, *Risk Communication, Risk Statistics, and Risk Comparisons: A Manual for Plant Managers* (Washington, DC: Chemical Manufacturers Association, 1988).

Edelstein, M. R., *Contaminated Communities: The Social and Psychological Impacts of Residential Toxic Exposure* (Boulder, CO: Westview Press, 1988).

Hance, B. J., C. Chess, and P. M. Sandman, *Improving Dialogue with Communities: A Risk Communication Manual for Government* (Trenton, NJ: Division of Science and Research, Department of Environmental Protection, 1988).

Kline, M., C. Chess, and P. M. Sandman, *Evaluating Risk Communication Programs: A Catalogue of "Quick and Easy" Feedback Methods* (New Brunswick, NJ: Environmental Communication Research Program, Rutgers University, 1989).

Krimsky, S., and A. Plough, *Environmental Hazards: Communicating Risks as a Social Process* (Dover, MA: Auburn House, 1988).

"Making Health Communication Programs Work: A Planner's Guide," Office of Cancer Communications, National Cancer Institute, Public Health Service, U.S. Department of Health and Human Services, Bethesda, MD (April 1989).

National Research Council, *Improving Risk Communication* (Washington, DC: National Academy Press, 1989).

Sandman, P. M., D. B. Sachsman, and M. R. Greenberg, *The Environmental News Source: Informing the Media During an Environmental Crisis* (Newark, NJ: Hazardous Substance Management Research Center, New Jersey Institute of Technology, 1987).

Science, Technology, and Human Values (special issue on risk communication), Vol. 12 (Issues 3 and 4), 1987.

INDICES OF GUIDELINES
AND CHECKLISTS

INDEX OF GUIDELINES

INDEX OF CHECKLISTS